BestMasters

Mit **„BestMasters“** zeichnet Springer die besten Masterarbeiten aus, die an renommierten Hochschulen in Deutschland, Österreich und der Schweiz entstanden sind. Die mit Höchstnote ausgezeichneten Arbeiten wurden durch Gutachter zur Veröffentlichung empfohlen und behandeln aktuelle Themen aus unterschiedlichen Fachgebieten der Naturwissenschaften, Psychologie, Sozialwissenschaften, Technik und Wirtschaftswissenschaften. Die Reihe wendet sich an Praktiker und Wissenschaftler gleichermaßen und soll insbesondere auch Nachwuchswissenschaftlern Orientierung geben.

Springer awards **“BestMasters”** to the best master’s theses which have been completed at renowned Universities in Germany, Austria, and Switzerland. The studies received highest marks and were recommended for publication by supervisors. They address current issues from various fields of research in natural sciences, psychology, social sciences, technology, and economics. The series addresses practitioners as well as scientists and, in particular, offers guidance for early stage researchers.

Tim-Jonas Peter

Reconstructing Wind Fields from Gravitational Data on Gas Giants

An Investigation of Mathematical Methods

Tim-Jonas Peter
Universität Siegen
Siegen, Germany

ISSN 2625-3577 ISSN 2625-3615 (electronic)
BestMasters
ISBN 978-3-658-51276-7 ISBN 978-3-658-51277-4 (eBook)
https://doi.org/10.1007/978-3-658-51277-4

This Springer Spektrum imprint is published by the registered company Springer Fachmedien Wiesbaden GmbH, part of Springer Nature.
The registered company address is: Abraham-Lincoln-Str. 46, 65189 Wiesbaden, Germany

Acknowledgements This work was produced in the context of a master thesis in mathematics at the University of Siegen. The supervisors were Prof. Dr. Volker Michel and Prof. Dr. Franz-Theo Suttmeier and their input was of great help while completing this thesis. Also of great help was the input provided by Dr. Johannes Wicht of the Max Planck Institute for Solar System Research in Göttingen, who provided the original impetus for the direction of this research. His feedback on the physical modelling was essential and the author is grateful for his cooperation.

Competing Interests The author has no competing interests to declare that are relevant to the content of this manuscript.

Contents

1 Introduction 1

2 Mathematical Background 3
- 2.1 Basic Definitions 3
- 2.2 Spherical Harmonics 7
- 2.3 The Newton Kernel and Gravitational Potential 9

3 Further Development of Solutions 11
- 3.1 Assumptions and Simplifications 11
- 3.2 Expansion of the Gravitational Potential 12
- 3.3 Solutions to the Helmholtz Equation 14

4 Modelling the Wind Inside the Planet 23
- 4.1 Simplifying the Euler Equation 24
- 4.2 Perturbation of the Thermal Wind Equation 26

5 The Static Model 29
- 5.1 Polytrope of Index Unity 29
- 5.2 The Case of Radial Symmetry 33

6 The simplest Form of the Thermal Wind Equation 37

7 The Thermo-Gravitational Wind Equation 41
- 7.1 Derivation of the Helmholtz Equation 41
- 7.2 An Orthonormal Basis for the Gravitational Potential 43
- 7.3 Determining the Wind-Induced Density 62

8 An Inverse Problem 67
8.1 Structure of the Wind Field 67
8.2 An Inverse Problem Involving the Wind-Induced Gravitational Coefficients 71

9 Conclusion 75

References 77

1 Introduction

A long standing subject of research in the planetary sciences has been the study of the interior dynamics of gas-giant-atmospheres, like those of Jupiter and Saturn. The surfaces of these planets can be observed by telescope—which has been going on since Galilei observed Jupiter and Saturn in the early 17th century—and today by spacecraft. They display a clearly visible banded structure which is caused by surface winds in the eastward and westward directions (see Fig. 8.1 in Sect. 8.1 of this thesis for some pictures). These wind fields are called zonal winds. The questions which many scientists have been trying to answer is if these banded structures in the wind field extend into the planetary interior, and if they do, in which manner and how far they extend down into the planet, as well as how strong these winds are at a specified depth (see for example [11, 19, 23]). To attempt to answer these questions further data related to the interior of Jupiter and Saturn is required. This data was provided by the Juno mission around Jupiter, which arrived in orbit around the planet in 2016, and the Cassini Grand Finale, which took place in 2017, for Saturn. These spacecrafts measured the gravitational field of their respective target planets, with the hope being that this data would provide information about mass transport caused by wind fields in the planetary interior. This would enable conclusions about the structure of the atmosphere and wind fields.

The problem of reconstructing planetary interiors from gravitational data has also been studied for Earth, where it is known as inverse gravimetry. Although the knowledge of the gravitational field is much more detailed in the case of Earth, it is known that the reconstruction of the mass density of a planet purely from this data is not possible, because the solutions of this inverse problem are not unique. See

---

Supplementary Information The online version contains supplementary material available at https://doi.org/10.1007/978-3-658-51277-4_1.

T. Peter, *Reconstructing Wind Fields from Gravitational Data on Gas Giants*, BestMasters, https://doi.org/10.1007/978-3-658-51277-4_1

[15] for a discussion of this. The aim of this thesis is to examine the mathematical methods used in the study of atmospherical dynamics based on gravitational data of gas giants and formulate them in a mathematically rigorous way. This will involve orthonormal bases for spaces of square-integrable functions, so we will start with some fundamental mathematics necessary in the study of these objects. After that we will examine some mathematical properties of the gravitational potential and of solutions to the Helmholtz equation. This preparation then allows us to describe the mathematical models which connect the zonal wind and the gravitational field. Finally we examine how this model can be used to formulate an inverse problem. This book also contains two appendices A and B, which are supplied in the electronic supplementary material (ESM). Appendix A deals with the calculation of the wind-induced gravitational coefficients δJ_n if the potential perturbation is known. Appendix B list some useful properties of ordinary and spherical Bessel functions, which are necessary to derive some of the statements in this work.

2 Mathematical Background

2.1 Basic Definitions

In this whole thesis $\mathbb{N}$ will refer to the natural numbers without zero and $\mathbb{N}_0 := \mathbb{N} \cup \{0\}$ will refer to the natural numbers with zero.
We begin by introducing some basic mathematical definitions about spherical geometry and regular surfaces. These definitions and theorems are taken from [13, 14], Chapter 2 in both books:

Definition 2.1.1 For some $R > 0$, let $B_R(0)$ denote the **open ball** of radius R around 0 in $\mathbb{R}^3$, meaning

$$B_R(0) := \{\boldsymbol{x} \in \mathbb{R}^3 \mid |\boldsymbol{x}| < R\} \subseteq \mathbb{R}^3.$$

Here $|\cdot|$ denotes the euclidean norm in $\mathbb{R}^3$. We also define the **unit sphere** $\mathbb{S}^2 := \partial B_1(0)$, specifically

$$\mathbb{S}^2 := \{\boldsymbol{x} \in \mathbb{R}^3 \mid |\boldsymbol{x}| = 1\}.$$

Finally it should be noted that every $\boldsymbol{x} \in \mathbb{R}^3 \setminus \{0\}$ can be uniquely decomposed as $\boldsymbol{x} = r\boldsymbol{\xi}$, where $r := |\boldsymbol{x}| \in \mathbb{R}$ and $\boldsymbol{\xi} := \frac{\boldsymbol{x}}{|\boldsymbol{x}|} \in \mathbb{S}^2$.

T. Peter, *Reconstructing Wind Fields from Gravitational Data on Gas Giants*, BestMasters, https://doi.org/10.1007/978-3-658-51277-4_2

Definition 2.1.2 Let $\Sigma \subseteq \mathbb{R}^3$ be a surface. Σ is called a **regular surface** if the following conditions are satisfied:

1. Σ subdivides $\mathbb{R}^3$ into a bounded region Σ_{int} (the interior) and an unbounded region Σ_{ext} (the exterior), such that $\mathbb{R}^3$ can be written as the disjoint union $\mathbb{R}^3 = \Sigma_{\text{ext}} \,\dot{\cup}\, \Sigma \,\dot{\cup}\, \Sigma_{\text{int}}$.
2. Σ is a closed (as in $\partial\Sigma = 0$) and compact surface, which is free of double points.
3. There exists a C^2-parametrization of Σ such that the Jacobian matrix of the parametrization has maximal rank$(= 2)$ in the interior of the parameter range.

Definition 2.1.3 We define the **standard basis vectors** $\boldsymbol{\varepsilon}^i \in \mathbb{R}^3$ for $i \in \{1, 2, 3\}$ by their components:

$$(\boldsymbol{\varepsilon}^i)_j = \delta_{ij} \text{ for } j \in \{1, 2, 3\}.$$

Here δ_{ij} is the Kronecker delta.

Definition 2.1.4 Every $\boldsymbol{x} \in \mathbb{R}^3$ can be written in **spherical coordinates** as

$$\boldsymbol{x} = \begin{pmatrix} r\sin(\theta)\cos(\varphi) \\ r\sin(\theta)\sin(\varphi) \\ r\cos(\theta) \end{pmatrix}$$

for $r = |\boldsymbol{x}| \geq 0, \theta \in [0, \pi]$ and $\varphi \in [0, 2\pi)$. Alternatively one can write

$$\boldsymbol{r} = \begin{pmatrix} r\sqrt{1-t^2}\cos(\varphi) \\ r\sqrt{1-t^2}\sin(\varphi) \\ rt \end{pmatrix}$$

with r and φ as above and $t(= \cos(\theta)) \in [-1, 1]$. We will prefer the second representation.

Definition 2.1.5 The vectors

$$\boldsymbol{\varepsilon}^r := \begin{pmatrix} \sqrt{1-t^2}\cos(\varphi) \\ \sqrt{1-t^2}\sin(\varphi) \\ t \end{pmatrix}, \ \boldsymbol{\varepsilon}^\varphi := \begin{pmatrix} -\sin(\varphi) \\ \cos(\varphi) \\ 0 \end{pmatrix} \ \text{and } \boldsymbol{\varepsilon}^t := \begin{pmatrix} -t\cos(\varphi) \\ -t\sin(\varphi) \\ \sqrt{1-t^2} \end{pmatrix}$$

with φ and t as above form a local orthonormal basis of $\mathbb{R}^3$, such that $\boldsymbol{\varepsilon}^r \times \boldsymbol{\varepsilon}^\varphi = \boldsymbol{\varepsilon}^t$. They will be referred to as **spherical basis vectors**. Also $\boldsymbol{\varepsilon}^r = \boldsymbol{\xi}$ with the $\boldsymbol{\xi}$ from Definition 2.1.1.

Definition 2.1.6 For a function $f \in \mathrm{C}^1(D)$, where $D \subseteq \mathbb{R}^3$ is open, we define the **surface gradient** ∇^* of f as the tangential part of the gradient, so

$$\nabla^* f(r\boldsymbol{\xi}(\varphi,t)) = \left(\sqrt{1-t^2}\boldsymbol{\varepsilon}^t\partial_t + \frac{1}{\sqrt{1-t^2}}\boldsymbol{\varepsilon}^\varphi\partial_\varphi\right) f(r\boldsymbol{\xi}(\varphi,t))$$

in spherical coordinates. We also sometimes write ∇^*_ξ to emphasize that this operator only includes derivatives tangential to the unit sphere.

Definition 2.1.7 For a function $f \in \mathrm{C}^1(D)$, where $D \subseteq \mathbb{R}^3$ is open, we define the **surface curl gradient** L^* of f as

$$\begin{aligned} \mathrm{L}^* f(r\boldsymbol{\xi}(\varphi,t)) &:= \boldsymbol{\xi} \times \nabla^* f(r\boldsymbol{\xi}(\varphi,t)) \\ &= \left(-\sqrt{1-t^2}\boldsymbol{\varepsilon}^\varphi\partial_t + \frac{1}{\sqrt{1-t^2}}\boldsymbol{\varepsilon}^t\partial_\varphi\right) f(r\boldsymbol{\xi}(\varphi,t)). \end{aligned}$$

Lemma 2.1.8 *Let $f \in \mathrm{C}^1(D)$, where $D \subseteq \mathbb{R}^3$ is open. Then the gradient of f at $D \ni \boldsymbol{x} \neq 0$ can be decomposed as*

$$\nabla_x f(\boldsymbol{x}) = \left(\boldsymbol{\xi}\partial_r + \frac{1}{r}\nabla^*_\xi\right) f(r\boldsymbol{\xi}).$$

Lemma 2.1.9 *For an open set $D \subseteq \mathbb{R}^3$ and a function $f \in \mathrm{C}^1(D)$, which is independent of the azimuthal angle φ in spherical coordinates (meaning that $\partial_\varphi f(r\boldsymbol{\xi}(\varphi,t)) = 0$) the surface curl gradient L^* can be expressed as*

$$\mathrm{L}^* f(r\boldsymbol{\xi}(\varphi,t)) = -\sqrt{1-t^2}\boldsymbol{\varepsilon}^\varphi\partial_t f(r\boldsymbol{\xi}(\varphi,t)).$$

This means L^* is proportional to $\boldsymbol{\varepsilon}^\varphi$.

Proof. Just use the fact that $\partial_\varphi f = 0$ and plug this into Definition 2.1.7. □

Definition 2.1.10 Δ will refer to the well-known **Laplace operator**. For a twice continuously differentiable function F it admits the following decomposition

$$\Delta F(\boldsymbol{x}) = \partial_r^2 F(r\boldsymbol{\xi}) + \frac{2}{r}\,\partial_r F(r\boldsymbol{\xi}) + \frac{1}{r^2}\,\Delta^* F(r\boldsymbol{\xi}) \quad \text{for every } \boldsymbol{x} \neq 0.$$

Here Δ^* is the **Laplace-Beltrami operator** which only acts on the coordinate $\boldsymbol{\xi} \in \mathbb{S}^2$. See Theorem 4.5 in [13] for further information.

We also need some L^2-function spaces.

Definition 2.1.11 For an open set $D \subseteq \mathbb{R}^3$ and two measurable functions $f, g : D \to \mathbb{R}$ we define the relation

$$f \sim g :\Leftrightarrow f(x) = g(x) \text{ for all } x \in D \setminus N,$$

where $N \subseteq D$ satisfies $\lambda(N) = 0$ (it has Lebesgue measure 0). The relation $\sim$ is obviously an **equivalence relation** and thus we refer to the **equivalence class** of the function f as $[f]$.

Remark 2.1.12 If $D \subseteq \mathbb{R}^3$ is open and $f : D \to \mathbb{R}$ is (Lebesgue-)integrable, then for any $g \in [f]$, the function g is also integrable and

$$\int_D f(\boldsymbol{x})\,\mathrm{d}\boldsymbol{x} = \int_D g(\boldsymbol{x})\,\mathrm{d}\boldsymbol{x}.$$

Definition 2.1.13 For $D \subseteq \mathbb{R}^3$ open, we define the **space of square-integrable functions** on D as the following set of equivalence classes of functions

$$\mathrm{L}^2(D) := \{[f] \mid f : D \to \mathbb{R} \text{ is measurable and } \|f\|_{\mathrm{L}^2(D)} < \infty\},$$

where the norm $\|\cdot\|_{\mathrm{L}^2(D)}$ is induced by the inner product

$$\langle [f], [g] \rangle_{\mathrm{L}^2(D)} := \int_D f(\boldsymbol{x}) g(\boldsymbol{x})\,\mathrm{d}\boldsymbol{x}.$$

It is well-known that this space is a Hilbert space. The inner product is independent of the representative of the equivalence class by Remark 2.1.12, therefore we will omit the equivalence class brackets from now on. The Hilbert space $\mathrm{L}^2(\mathbb{S}^2)$ is defined completely analogously but here the inner product is

$$\langle f, g\rangle_{\mathrm{L}^2(\mathbb{S}^2)} := \int_{\mathbb{S}^2} f(\xi)g(\xi)\,\mathrm{d}\omega(\xi),$$

where $\mathrm{d}\omega$ denotes the surface measure on $\mathbb{S}^2$ induced by the Lebesgue measure. We will also need weighted L^2-spaces on an interval $(a, b) \neq \emptyset$ for a weight function w, which is positive on (a, b) and continuous on $[a, b]$. This space is defined analogously to the ones above but our inner product is

$$\langle u, v\rangle_{\mathrm{L}^2_w(a,b)} = \int_a^b w(r)u(r)v(r)\,\mathrm{d}r.$$

We will refer to this space by $\mathrm{L}^2_w(a, b)$.

2.2 Spherical Harmonics

This section is adapted from [13] Chap. 5, the proofs for the theorems and lemmata stated here can be found in this book.

Definition 2.2.1 The **Legendre polynomials** P_n for $n \in \mathbb{N}_0$ form the unique sequence of polynomials on $[-1, 1]$, such that

1. $\deg(P_n) = n$ for all $n \in \mathbb{N}_0$,
2. If $n \neq m$ we have $\int_{-1}^1 P_n(t)P_m(t)\,\mathrm{d}t = 0$ and
3. $P_n(1) = 1$ for all $n \in \mathbb{N}_0$.

It can be shown that these three conditions determine the sequence $(P_n)_n$ exactly (see for example [13] Theorem 3.9 and Definition 3.10).

Definition 2.2.2 For $n \in \mathbb{N}_0$ and $m \in \{0, \ldots, n\}$, we define

$$P_{n,m}(t) := (1 - t^2)^{m/2}\frac{\mathrm{d}^m}{\mathrm{d}t^m}P_n(t) \quad \text{for } t \in [-1, 1]$$

as the **associated Legendre functions**. Note that $P_{n,0}(t) = P_n(t)$.

Definition 2.2.3 A polynomial in $\mathbb{R}^3$ of the form

$$P(x) = \sum_{|\alpha|=m} C_\alpha x^\alpha,$$

where $\alpha \in \mathbb{N}_0^3$ and not all of the coefficients $C_\alpha \in \mathbb{R}$ are zero, is called homogeneous of degree $m \in \mathbb{N}_0$. For a fixed value of m, the space of all these polynomials in addition to the 0-polynomial is called $\mathrm{Hom}_m(\mathbb{R}^3)$. The space of **homogeneous polynomials which are also harmonic** is denoted as

$$\mathrm{Harm}_m(\mathbb{R}^3) := \{P \in \mathrm{Hom}_m(\mathbb{R}^3) \mid \Delta P = 0 \text{ on } \mathbb{R}^3\}.$$

Definition 2.2.4 We define the space of **spherical harmonics** of fixed degree $n \in \mathbb{N}_0$ as

$$\mathrm{Harm}_n(\mathbb{S}^2) := \{P|_{\mathbb{S}^2} \mid P \in \mathrm{Harm}_n(\mathbb{R}^3)\}.$$

The elements $Y_n \in \mathrm{Harm}_n(\mathbb{S}^2)$ are called spherical harmonics. It is known that $\dim(\mathrm{Harm}_n(\mathbb{S}^2)) = 2n + 1$ (see [13] Theorem 5.6), so we will denote any $\mathrm{L}^2(\mathbb{S}^2)$-orthonormal basis of $\mathrm{Harm}_n(\mathbb{S}^2)$ by $(Y_{n,j})_{j=-n,\ldots,n}$.

Lemma 2.2.5 *For a fixed but arbitrary $n \in \mathbb{N}_0$, every $Y_n \in \mathrm{Harm}_n(\mathbb{S}^2)$ satisfies*

$$\Delta^* Y_n(\xi) = -n(n+1)Y_n(\xi) \quad \textit{for every } \xi \in \mathbb{S}^2.$$

Lemma 2.2.6 *For $m, k \in \mathbb{N}_0$ with $m \neq k$ and $Y_m \in \mathrm{Harm}_m(\mathbb{S}^2)$ as well as $Y_k \in \mathrm{Harm}_k(\mathbb{S}^2)$, we have*

$$\langle Y_m, Y_k \rangle_{\mathrm{L}^2(\mathbb{S}^2)} = 0.$$

Theorem 2.2.7 *For any $\mathrm{L}^2(\mathbb{S}^2)$-orthonormal system $(Y_{n,j})_{j=-n,\ldots,n}$ of $\mathrm{Harm}_n(\mathbb{S}^2)$ (here $n \in \mathbb{N}_0$ is fixed), the addition theorem*

$$\sum_{j=-n}^{n} Y_{n,j}(\xi)Y_{n,j}(\zeta) = \frac{2n+1}{4\pi} P_n(\xi \cdot \zeta) \quad \textit{for any } \xi, \zeta \in \mathbb{S}^2$$

holds, where P_n is the n-th Legendre polynomial.

Lemma 2.2.8 *Every* $Y_n \in \mathrm{Harm}_n(\mathbb{S}^2)$ *satisfies*

$$\|Y_n\|_{\mathrm{C}(\mathbb{S}^2)} \leq \sqrt{\frac{2n+1}{4\pi}} \|Y_n\|_{\mathrm{L}^2(\mathbb{S}^2)}.$$

Theorem 2.2.9 *A* $\mathrm{L}^2(\mathbb{S}^2)$*-orthonormal system* $(Y_{n,j})_{n,j}$ *(where* $n \in \mathbb{N}_0$ *and* $j \in \{-n, \ldots, n\}$*) of spherical harmonics is complete in* $\mathrm{L}^2(\mathbb{S}^2)$*, meaning that for any* $F \in \mathrm{L}^2(\mathbb{S}^2)$*, we have*

$$\left\| F - \sum_{n=0}^{N} \sum_{j=-n}^{n} \langle F, Y_{n,j} \rangle_{\mathrm{L}^2(\mathbb{S}^2)} Y_{n,j} \right\|_{\mathrm{L}^2(\mathbb{S}^2)} \xrightarrow[N\to\infty]{} 0.$$

Definition 2.2.10 For a vector $\xi(\varphi, t) \in \mathbb{S}^2$, given in spherical coordinates, we define the **fully normalized spherical harmonics** $Y_{n,j} : \mathbb{S}^2 \to \mathbb{R}$ as

$$Y_{n,j}(\xi(\varphi, t)) := \sqrt{\frac{(2n+1)(n-|j|)!(2-\delta_{j0})}{4\pi(n+|j|)!}} P_{n,|j|}(t) \begin{cases} \sin(j\varphi) & \text{for } j > 0 \\ \cos(j\varphi) & \text{for } j \leq 0, \end{cases}$$

where $n \in \mathbb{N}_0$ and $j \in \{-n, \ldots, n\}$. Here δ_{j0} denotes the usual Kronecker delta. The name is justified because these functions are indeed in $\mathrm{Harm}_n(\mathbb{S}^2)$ and $\mathrm{L}^2(\mathbb{S}^2)$-orthonormal, thus all the properties of the above section apply (see Sect. 5.2 in [13]). From now on $Y_{n,j}$ will refer to the fully normalized spherical harmonics.

2.3 The Newton Kernel and Gravitational Potential

For the problem of inverse gravimetry we need to define and study the properties of the Newton kernel.

Definition 2.3.1 For $\boldsymbol{x}, \boldsymbol{y} \in \mathbb{R}^3$ with $\boldsymbol{x} \neq \boldsymbol{y}$ we define the **Newton kernel** as

$$N(\boldsymbol{x} - \boldsymbol{y}) := \frac{1}{|\boldsymbol{x} - \boldsymbol{y}|}.$$

Definition 2.3.2 Let Σ be a regular surface, $B := \Sigma_{\text{int}}$ and $F \in \mathrm{C}^2(\overline{B})$ a mass density. We define the **gravitational potential** V caused by F as

$$V(\boldsymbol{x}) := -G \int_B \frac{F(\boldsymbol{y})}{|\boldsymbol{x} - \boldsymbol{y}|} \,\mathrm{d}\boldsymbol{y}$$

for any $\boldsymbol{x} \in \mathbb{R}^3$. G is the gravitational constant ($G \approx 6.67 \times 10^{-11}\ \mathrm{Nm^2kg^{-2}}$).

Definition 2.3.3 A function $V \in \mathrm{C}^2(\mathbb{R}^3 \setminus \overline{B})$, where B is the interior of some regular surface, is called **regular at infinity** if

$$\begin{aligned} |V(y)| &= \mathcal{O}\left(|y|^{-1}\right) \quad \text{for } |y| \to \infty \quad \text{and} \\ |\nabla V(y)| &= \mathcal{O}\left(|y|^{-2}\right) \quad \text{for } |y| \to \infty. \end{aligned}$$

We now summarize some basic facts about the gravitational potential as defined above (For proofs of these statements see the summary at the end of Sect. 3.1 in [14]):

1. V exists on all of $\mathbb{R}^3$, is bounded, continuous and partially differentiable everywhere. We can also exchange integration and differentiation.
2. $V \in \mathrm{C}^2(\mathbb{R}^3 \setminus \overline{B})$ and satisfies $\Delta V = 0$ on $\mathbb{R}^3 \setminus \overline{B}$. V is also regular at infinity.
3. $V \in \mathrm{C}^2(B)$ and satisfies $\Delta V = 4\pi G F$ on B.

It is also worth noting that V is continuously differentiable everywhere. Although this is not proven in [14], a proof of this is easily adapted from Theorem 3.1.2 in [14], using Corollary 3.1.4 in the same book. Finally we note one important property of the Newton kernel.

Theorem 2.3.4 *For any* $\boldsymbol{x}, \boldsymbol{y} \in \mathbb{R}^3$ *with* $|\boldsymbol{x}| < |\boldsymbol{y}|$, *the identity*

$$\frac{1}{|\boldsymbol{x} - \boldsymbol{y}|} = \sum_{n=0}^{\infty} \frac{|\boldsymbol{x}|^n}{|\boldsymbol{y}|^{n+1}} P_n\left(\frac{\boldsymbol{x}}{|\boldsymbol{x}|} \cdot \frac{\boldsymbol{y}}{|\boldsymbol{y}|}\right)$$

holds.

Proof. See Corollary 3.4.18 in [14]. □

3 Further Development of Solutions

We model our gas giant as an open set $B \subseteq \mathbb{R}^3$, where $B := \Sigma_{\text{int}}$ is the interior of a regular surface Σ. The origin of our coordinate system is the center of mass of the planet. The simplest model is $B = B_R(0)$, with the equatorial radius $R > 0$.

3.1 Assumptions and Simplifications

Assumption 3.1.1 By observation of gas giants like Jupiter and Saturn, it is clear that these planets are symmetric under rotation about the axis of rotation [8, 9]. This means all ocurring physical quantities Q **cannot depend on the azimuthal angle** $\boldsymbol{\varphi}$ in spherical coordinates, so $\partial_\varphi Q(r\boldsymbol{\xi}(\varphi, t)) = 0$ for all r, φ and t.

This simplification allows us to state the following lemma.

Lemma 3.1.2 *For* $n \in \mathbb{N}$, $j \in \{-n, \ldots, n\}$ *with* $j \neq 0$ *and a mass density* $F \in \mathrm{C}^2(\overline{B})$ *satisfying Assumption 3.1.1, we have*

$$\int_B |\boldsymbol{x}|^n Y_{n,j}\left(\frac{\boldsymbol{x}}{|\boldsymbol{x}|}\right) F(\boldsymbol{x})\,\mathrm{d}\boldsymbol{x} = 0.$$

Proof. If we use spherical coordinates to calculate the above integral, then because F does not depend on φ, the integrals over the coordinate φ look like

$$\int_0^{2\pi} \sin(j\varphi)\,\mathrm{d}\varphi \quad \text{or} \quad \int_0^{2\pi} \cos(j\varphi)\,\mathrm{d}\varphi,$$

T. Peter, *Reconstructing Wind Fields from Gravitational Data on Gas Giants*, BestMasters, https://doi.org/10.1007/978-3-658-51277-4_3

which both vanish because $j \neq 0$. This means that the whole integral must vanish. □

Lemma 3.1.3 *Let $F \in \mathrm{C}^2(\overline{B})$ be the mass density of our planet B with total mass $M := \int_B F(\boldsymbol{x})\,\mathrm{d}\boldsymbol{x} > 0$. Then the center of mass is*

$$\boldsymbol{x}_M := \frac{1}{M} \int_B \boldsymbol{x} F(\boldsymbol{x})\,\mathrm{d}\boldsymbol{x}$$

and we can conlcude that

$$\int_B x_3 F(\boldsymbol{x})\,\mathrm{d}\boldsymbol{x} = 0.$$

Proof. $\boldsymbol{x}_M = 0$, since the center of mass is at the origin. Then the integral is just the third component of $M\boldsymbol{x}_M = 0$. □

3.2 Expansion of the Gravitational Potential

We now have every tool to expand the gravitational potential of a planet modeled as $B = B_R(0)$ on $\mathbb{R}^3 \setminus B$ into spherical harmonics.

Theorem 3.2.1 *If $F \in \mathrm{C}^2(\overline{B_R(0)})$ is a mass density satisfying Assumption 3.1.1, then its corresponding gravitational potential can be written as*

$$\begin{aligned} V(r\boldsymbol{\xi}(\varphi, t)) &= -\frac{GM}{r}\left(1 - \sum_{n=2}^{\infty} J_n \left(\frac{R}{r}\right)^n P_n(\xi_3)\right) \\ &= -\frac{GM}{r}\left(1 - \sum_{n=2}^{\infty} J_n \left(\frac{R}{r}\right)^n P_n(t)\right), \end{aligned}$$

if $r > R$. The coefficients J_n are defined as

$$J_n := -\frac{1}{MR^n} \int_0^R \int_{\mathbb{S}^2} r^{n+2} F(r\boldsymbol{\xi}) P_n(\xi_3)\,\mathrm{d}\omega(\boldsymbol{\xi})\,\mathrm{d}r \text{ for } n \geq 2.$$

Proof. The first steps are to use Theorem 2.3.4 on the Newton kernel in the integral (which we can do because we assume $r > R$), then interchange series and integral, which is justified by the convergence of the integral in $L^2(B)$, and lastly use Theorem 2.2.7.

$$
\begin{aligned}
V(r\xi) =& - G \int_{B_R(0)} \frac{F(y)}{|r\xi - y|} \,dy = -G \int_{\mathbb{S}^2} \int_0^R \frac{F(s\zeta)}{|s\zeta - r\xi|} s^2 \,ds\, d\omega(\zeta) \\
=& - G \sum_{n=0}^{\infty} \int_{\mathbb{S}^2} \int_0^R \frac{s^n}{r^{n+1}} F(s\zeta) P_n(\zeta \cdot \xi) s^2 \,ds\, d\omega(\zeta) \\
=& - G \sum_{n=0}^{\infty} \sum_{j=-n}^{n} r^{-n-1} \sqrt{\frac{4\pi}{2n+1}} Y_{n,j}(\xi) \\
&\times \int_{\mathbb{S}^2} \int_0^R s^{n+2} \sqrt{\frac{4\pi}{2n+1}} Y_{n,j}(\zeta) F(s\zeta) \,ds\, d\omega(\zeta).
\end{aligned}
$$

Now we define the coefficients $V_{n,j}$ for $n \in \mathbb{N}_0$ and $j \in \{-n, \ldots, n\}$ by

$$
\begin{aligned}
V_{n,j} :=& \sqrt{\frac{4\pi}{2n+1}} \int_0^R \int_{\mathbb{S}^2} r^{n+2} Y_{n,j}(\xi) F(r\xi) \,dr\, d\omega(\xi) \\
=& \sqrt{\frac{4\pi}{2n+1}} \int_B |x|^n Y_{n,j}\left(\frac{x}{|x|}\right) F(x) \,dx.
\end{aligned}
$$

Since F satisfies Assumption 3.1.1, we conclude by Lemma 3.1.2 that $V_{n,j} = 0$ for $n \in \mathbb{N}$ and $j \neq 0$. If we look at Definition 2.2.10, we see that

$$
Y_{n,0}(\xi) = \sqrt{\frac{2n+1}{4\pi}} P_n(\xi_3)
$$

($\xi_3 = t$ in spherical coordinates). Since $P_0(t) = 1$ and $P_1(t) = t$ we have

$$
\begin{aligned}
V_{0,0} &= \int_B F(x) \,dx = M, \\
V_{1,0} &= \int_B x_3 F(x) \,dx = 0 \text{ by Lemma 3.1.3 and} \\
V_{n,0} &= -M R^n J_n \text{ for } n \geq 2.
\end{aligned}
$$

Now we can plug this all back into the expansion of V.

$$
\begin{aligned}
V(r\xi) &= -G \sum_{n=0}^{\infty} V_{n,0} r^{-n-1} P_n(\xi_3) \\
&= -\frac{GM}{r}\left(1 + \sum_{n=2}^{\infty}\left(\frac{R}{r}\right)^n \frac{1}{MR^n} V_{n,0} P_n(\xi_3)\right) \\
&= -\frac{GM}{r}\left(1 - \sum_{n=2}^{\infty} J_n \left(\frac{R}{r}\right)^n P_n(\xi_3)\right),
\end{aligned}
$$

this proves the theorem. □

This expansion is common in the literature and the coefficients J_n, calculated from the observed data, are available for computations. See for example [10], for a table containing the coefficients J_2 to J_{40}.

3.3 Solutions to the Helmholtz Equation

In this thesis we will sometimes derive a **Helmholtz equation**

$$
(\Delta + \gamma^2)\, h(\boldsymbol{x}) = 0 \text{ on } B \subseteq \mathbb{R}^3 \tag{3.1}
$$

for some quantity $h \in C^2(B)$ (which will either be a mass density or a potential) and $\gamma > 0$. We require that B is a bounded and open set, such that $0 \in B$. In the following Γ will refer to the well-known gamma-function. The basic definitions are from [1] 9.1.10 and 10.1.1, the rest of the chapter is adapted from [16] pages 82–83.

Definition 3.3.1 The **Bessel function** J_ν of the first kind and of order $\nu \in \mathbb{R}$ is defined via the following power series. For $x \in \mathbb{R}$ we have

$$
J_\nu(x) := \sum_{m=0}^{\infty} \frac{(-1)^m}{m!\,\Gamma(\nu + m + 1)} \left(\frac{x}{2}\right)^{2m+\nu}.
$$

Definition 3.3.2 The **spherical Bessel functions** of the first kind and of order $l \in \mathbb{Z}$ are defined via

$$
j_l(x) := \sqrt{\frac{\pi}{2x}}\, J_{l+\frac{1}{2}}(x) \quad \text{for } x > 0.
$$

Lemma 3.3.3 *For any $D > 0$, the sequence*

$$\left(\frac{2j_n(x)\Gamma\left(n+\frac{3}{2}\right)}{\left(\frac{x}{2}\right)^n\Gamma\left(\frac{1}{2}\right)}\right)_{n\in\mathbb{N}_0}$$

converges to 1 for $n \to \infty$, uniformly in $x \in [0, D]$.
This means that the spherical Bessel functions $j_n(x)$ behave as

$$\frac{1}{2}\left(\frac{x}{2}\right)^n\frac{\Gamma\left(\frac{1}{2}\right)}{\Gamma\left(n+\frac{3}{2}\right)}$$

asymptotically for $n \to \infty$, uniformly in $x \in [0, D]$.

Proof. With the power series expression for $J_{n+\frac{1}{2}}$ and the definition of the spherical Bessel functions we get a power series expression for $j_n(x)$, namely

$$j_n(x) = \frac{\Gamma\left(\frac{1}{2}\right)}{2}\sum_{m=0}^{\infty}\frac{(-1)^m}{m!\,\Gamma\left(m+n+\frac{3}{2}\right)}\left(\frac{x}{2}\right)^{2m+n},$$

where we have used $\Gamma\left(\frac{1}{2}\right) = \sqrt{\pi}$. This means the sequence we want to look at has the form

$$\begin{aligned}\frac{2j_n(x)\Gamma\left(n+\frac{3}{2}\right)}{\left(\frac{x}{2}\right)^n\Gamma\left(\frac{1}{2}\right)} &= \sum_{m=0}^{\infty}\frac{(-1)^m\Gamma\left(n+\frac{3}{2}\right)}{m!\,\Gamma\left(m+n+\frac{3}{2}\right)}\left(\frac{x}{2}\right)^{2m}\\ &= 1+\sum_{m=1}^{\infty}\frac{(-1)^m\Gamma\left(n+\frac{3}{2}\right)}{m!\,\Gamma\left(m+n+\frac{3}{2}\right)}\left(\frac{x}{2}\right)^{2m}.\end{aligned}$$

Because $\Gamma(z+1) = z\Gamma(z)$ we can simplify in the denominator

$$\begin{aligned}\Gamma\left(m+n+\frac{3}{2}\right) &= \Gamma\left(n+\frac{3}{2}\right)\prod_{k=0}^{m-1}\left(k+n+\frac{3}{2}\right)\\ &= \left(n+\frac{3}{2}\right)\Gamma\left(n+\frac{3}{2}\right)\prod_{k=1}^{m-1}\left(k+n+\frac{3}{2}\right).\end{aligned}$$

Therefore we can estimate a coefficient occuring in the above power series by

$$\frac{\Gamma\left(n+\frac{3}{2}\right)}{\Gamma\left(m+n+\frac{3}{2}\right)} = \frac{1}{n+\frac{3}{2}} \prod_{k=1}^{m-1} \frac{1}{k+n+\frac{3}{2}} \leq \frac{1}{n+\frac{3}{2}} \prod_{k=1}^{m-1} \frac{1}{k}$$
$$= \frac{1}{n+\frac{3}{2}} \frac{1}{(m-1)!}.$$

With this estimate we can calculate the error of the sequence we are looking at

$$\left| \frac{2 j_n(x) \Gamma\left(n+\frac{3}{2}\right)}{\left(\frac{x}{2}\right)^n \Gamma\left(\frac{1}{2}\right)} - 1 \right| \leq \frac{1}{n+\frac{3}{2}} \sum_{m=1}^{\infty} \frac{1}{m!\,(m-1)!} \left(\frac{x}{2}\right)^{2m}$$
$$\leq \frac{1}{n+\frac{3}{2}} \sum_{m=1}^{\infty} \frac{1}{m!\,(m-1)!} \left(\frac{D}{2}\right)^{2m}.$$

We can easily check that the series on the right-hand side converges, thus we can estimate the error by C/n for some constant $C > 0$, which only depends on D. This proves the uniform convergence. □

We want to prove that we can express a solution h to the Helmholtz equation (3.1) as a series of spherical harmonics and spherical Bessel functions. For this we need a couple of lemmata.

Lemma 3.3.4 *If* $(Y_n)_{n \in \mathbb{N}_0} \subseteq \mathrm{Harm}_n(\mathbb{S}^2)$, $C_n > 0$ *with*

$$C_n^2 := \int_{\mathbb{S}^2} |Y_n(\xi)|^2 \, d\omega(\xi)$$

and the series

$$\sum_{n=0}^{\infty} |j_n(\gamma r_0)|^2 C_n^2$$

converges for some $r_0 > 0$, *then the series*

$$\sum_{n=0}^{\infty} j_n(\gamma r) Y_n(\xi)$$

converges for all $0 \leq r < r_0$. *This convergence is absolute and uniform in every closed subset of* $B_{r_0}(0)$.

Proof. The convergence of the series gets us

$$|j_n(\gamma r_0)|^2 C_n^2 \xrightarrow[n\to\infty]{} 0,$$

meaning this sequence is bounded. Because of Lemma 3.3.3, there is a $N_0 \in \mathbb{N}$ such that $j_n(\gamma r_0) \neq 0$ for all $n \geq N_0$. These two facts imply that there is a constant $A > 0$ (which depends on r_0) and $N_1(r_0) \in \mathbb{N}$ with

$$C_n^2 \leq \frac{A^2}{|j_n(\gamma r_0)|^2} \quad \text{for all } n \geq N_1(r_0).$$

Now Lemma 2.2.8 yields $|Y_n(\xi)| \leq \sqrt{(2n+1)/(4\pi)}C_n$ for all $\xi \in \mathbb{S}^2$, meaning

$$|j_n(\gamma r)Y_n(\xi)| \leq C_n |j_n(\gamma r)| \sqrt{\frac{2n+1}{4\pi}} \leq A\sqrt{\frac{2n+1}{4\pi}} \left|\frac{j_n(\gamma r)}{j_n(\gamma r_0)}\right|$$

for all $n \geq N_2(r_0) := \max\{N_0, N_1(r_0)\}$. Using Lemma 3.3.3 once again gets us that

$$\lim_{n\to\infty} \frac{r_0^n j_n(\gamma r)}{r^n j_n(\gamma r_0)} = 1 \quad \text{uniformly in } r \in [0, r_0].$$

Therefore there is a natural number $N_3(r_0)$, such that $|j_n(\gamma r)/j_n(\gamma r_0)| \leq 2(r/r_0)^n$ for every $n \geq N_3(r_0)$. Putting this all together, we gain the existence of a natural number $N_4(r_0) := \max\{N_2(r_0), N_3(r_0)\}$, such that for every $n \geq N_4(r_0)$ we have

$$|j_n(\gamma r)Y_n(\xi)| \leq A\sqrt{\frac{2n+1}{\pi}} \left(\frac{r}{r_0}\right)^n.$$

So we have estimated the term we want to sum over by a geometric series. This proves the statement. □

Lemma 3.3.5 *Let $f, g : \mathbb{S}^2 \to \mathbb{R}$ such that $f, g \in \mathrm{C}^2(\mathbb{S}^2)$. Then*

$$\int_{\mathbb{S}^2} f(\xi)\Delta^* g(\xi)\,\mathrm{d}\omega(\xi) = \int_{\mathbb{S}^2} g(\xi)\Delta^* f(\xi)\,\mathrm{d}\omega(\xi).$$

Proof. We know that $\nabla^* \cdot \nabla^* = \Delta^*$ (see Theorem 4.7 in [13]). Then by the product rule we have

$$\int_{\mathbb{S}^2} \nabla^* \cdot \big(f(\boldsymbol{\xi})\nabla^* g(\boldsymbol{\xi})\big)\, \mathrm{d}\omega(\boldsymbol{\xi}) = \int_{\mathbb{S}^2} \nabla^* f(\boldsymbol{\xi}) \cdot \nabla^* g(\boldsymbol{\xi})\, \mathrm{d}\omega(\boldsymbol{\xi}) + \int_{\mathbb{S}^2} f(\boldsymbol{\xi})\Delta^* g(\boldsymbol{\xi})\, \mathrm{d}\omega(\boldsymbol{\xi}).$$

We need to look at the left-hand side of this equation more carefully. For this, let $h : \mathbb{S}^2 \to \mathbb{R}^3$ be a C^1-vector field on $\mathbb{S}^2$ and define the C^1-vector field $H : \overline{B_1(0)} \setminus \{0\} \to \mathbb{R}^3$, $r\boldsymbol{\xi} \mapsto h(\boldsymbol{\xi})$. For every $0 < \varepsilon < 1$ we can apply Gauß's Law to H on the set $D_\varepsilon := B_1(0) \setminus B_\varepsilon(0)$ to get

$$\int_{D_\varepsilon} \nabla \cdot H(\boldsymbol{x})\, \mathrm{d}\boldsymbol{x} = \int_{\partial D_\varepsilon} \boldsymbol{n} \cdot H(\boldsymbol{x})\, \mathrm{dS}(\boldsymbol{x}). \tag{3.2}$$

The boundary of D_ε consists of two disjoint sets, the sphere of radius 1 and the sphere of radius ε around 0. On the first set, the outer normal is $\boldsymbol{n} = \boldsymbol{\xi}$ and on the second set the outer normal is $\boldsymbol{n} = -\boldsymbol{\xi}$. This is illustrated for the two-dimensional case in Figure 3.1. We can now evaluate the boundary integral in the limit $\varepsilon \to 0$ via

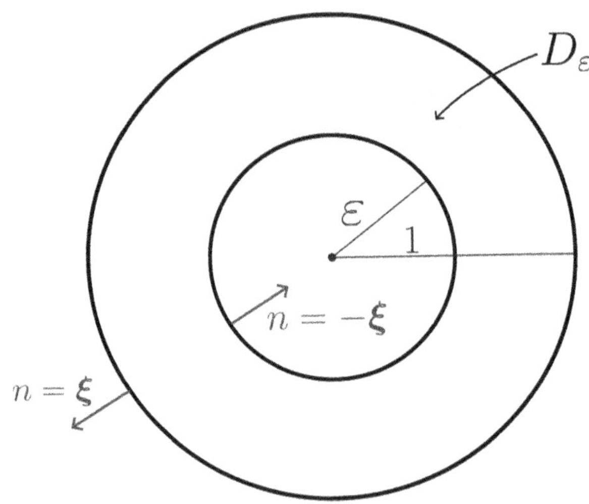

Fig. 3.1 The domain of the integral in two dimensions with outer normals

$$\begin{aligned}&\left|\int_{\partial D_\varepsilon} \boldsymbol{n}\cdot H(\boldsymbol{x})\,\mathrm{d}S(\boldsymbol{x}) - \int_{\mathbb{S}^2} \boldsymbol{\xi}\cdot H(\boldsymbol{\xi})\,\mathrm{d}\omega(\boldsymbol{\xi})\right| = \varepsilon^2\left|\int_{\mathbb{S}^2} \boldsymbol{\xi}\cdot H(\varepsilon\boldsymbol{\xi})\,\mathrm{d}\omega(\boldsymbol{\xi})\right|\\
\leq{}& \varepsilon^2\int_{\mathbb{S}^2} \|\,|H|\,\|_{\mathrm{C}(\overline{D_\varepsilon})}\,\mathrm{d}\omega(\boldsymbol{\xi}) = 4\pi\varepsilon^2\|\,|h|\,\|_{\mathrm{C}(\mathbb{S}^2)} \xrightarrow[\varepsilon\to 0]{} 0.\end{aligned}$$

Here $|H|$ is the function which maps $\boldsymbol{x}$ to $|H(\boldsymbol{x})|$. We can also evaluate the volume integral of Gauß's Law for $\varepsilon \to 0$, by noticing that $\partial_r H(r\boldsymbol{\xi}) = 0$. Therefore $\nabla\cdot H(r\boldsymbol{\xi}) = \nabla^*\cdot h(\boldsymbol{\xi})/r$ for every $r\boldsymbol{\xi} \in B_\varepsilon(0)$ and

$$\begin{aligned}&\left|\int_{D_\varepsilon} \nabla\cdot H(\boldsymbol{x})\,\mathrm{d}\boldsymbol{x} - \int_{B_1(0)} \nabla\cdot H(\boldsymbol{x})\,\mathrm{d}\boldsymbol{x}\right| = \left|\int_{B_\varepsilon(0)} \nabla\cdot H(\boldsymbol{x})\,\mathrm{d}\boldsymbol{x}\right|\\
={}&\left|\int_0^\varepsilon r^2\int_{\mathbb{S}^2}\frac{\nabla^*\cdot h(\boldsymbol{\xi})}{r}\,\mathrm{d}\omega(\boldsymbol{\xi})\,\mathrm{d}r\right| = \left|\int_0^\varepsilon r\,\mathrm{d}r\right|\left|\int_{\mathbb{S}^2}\nabla^*\cdot h(\boldsymbol{\xi})\,\mathrm{d}\omega(\boldsymbol{\xi})\right|\\
\leq{}&\frac{\varepsilon^2}{2}\int_{\mathbb{S}^2}\|\nabla^*\cdot h\|_{\mathrm{C}(\mathbb{S}^2)}\,\mathrm{d}\omega(\boldsymbol{\xi}) = 2\pi\varepsilon^2\|\nabla^*\cdot h\|_{\mathrm{C}(\mathbb{S}^2)}\xrightarrow[\varepsilon\to 0]{} 0.\end{aligned}$$

Taking the limit $\varepsilon \to 0$ on both sides of Equation (3.2) has led us to

$$\int_{B_1(0)} \nabla\cdot H(\boldsymbol{x})\,\mathrm{d}\boldsymbol{x} = \int_{\mathbb{S}^2} \boldsymbol{\xi}\cdot H(\boldsymbol{\xi})\,\mathrm{d}\omega(\boldsymbol{\xi}).$$

Applying $\nabla\cdot H(r\boldsymbol{\xi}) = \nabla^* h(\boldsymbol{\xi})/r$ again and recognizing that $H|_{\mathbb{S}^2} = h$, the previous equation is equivalent to

$$\begin{aligned}&\int_0^1 r\int_{\mathbb{S}^2}\nabla^*\cdot h(\boldsymbol{\xi})\,\mathrm{d}\omega(\boldsymbol{\xi})\,\mathrm{d}r = \int_{\mathbb{S}^2}\boldsymbol{\xi}\cdot h(\boldsymbol{\xi})\,\mathrm{d}\omega(\boldsymbol{\xi})\\
&\qquad\Leftrightarrow \int_{\mathbb{S}^2}\nabla^*\cdot h(\boldsymbol{\xi})\,\mathrm{d}\omega(\boldsymbol{\xi}) = 2\int_{\mathbb{S}^2}\boldsymbol{\xi}\cdot h(\boldsymbol{\xi})\,\mathrm{d}\omega(\boldsymbol{\xi}).\end{aligned}$$

Using this fact, we conclude that

$$\int_{\mathbb{S}^2}\nabla^*\cdot\big(f(\boldsymbol{\xi})\nabla^* g(\boldsymbol{\xi})\big)\,\mathrm{d}\omega(\boldsymbol{\xi}) = 2\int_{\mathbb{S}^2} f(\boldsymbol{\xi})\big(\boldsymbol{\xi}\cdot\nabla^* g(\boldsymbol{\xi})\big)\,\mathrm{d}\omega(\boldsymbol{\xi}) = 0,$$

because $\boldsymbol{\xi}\cdot\nabla^* = 0$. Therefore

$$\int_{\mathbb{S}^2} f(\boldsymbol{\xi})\Delta^* g(\boldsymbol{\xi})\,\mathrm{d}\omega(\boldsymbol{\xi}) = -\int_{\mathbb{S}^2}\nabla^* f(\boldsymbol{\xi})\cdot\nabla^* g(\boldsymbol{\xi})\,\mathrm{d}\omega(\boldsymbol{\xi}).$$

If we now switch f and g then, because the right-hand side of the above equation is symmetric in f and g, we get the desired relation. □

Lemma 3.3.6 *If $h \in \mathrm{C}^2(B_R(0))$ satisfies Equation* (3.1), *then the functions*

$$H_{n,j}(r) := \int_{\mathbb{S}^2} h(r\xi)Y_{n,j}(\xi)\,\mathrm{d}\omega(\xi)$$
$$\text{for } r \in [0,R) \text{ and } n \in \mathbb{N}_0,\ j \in \{-n,\ldots,n\}$$

are in $\mathrm{C}^2[0,R)$ and satisfy the differential equation

$$H''_{n,j}(r) + \frac{2}{r}H'_{n,j}(r) - \frac{n(n+1)}{r^2}H_{n,j}(r) = -\gamma^2 H_{n,j}(r)$$

on $(0,R)$. This means that the $H_{n,j}$ have the form

$$H_{n,j}(r) = c_{n,j}\, j_n(\gamma r) \quad \text{for } r \in [0,R) \text{ and } n \in \mathbb{N}_0,\ j \in \{-n,\ldots,n\},$$

where the $c_{n,j}$ are real constants.

Proof. Since h is in $\mathrm{C}^2(B_R(0))$, the function $H_{n,j}$ is twice continuously differentiable on its domain as well. Also since the integrand which defines the $H_{n,j}$ is in C^2 and the integration domain is compact, we can switch differentiation with respect to r and integration. Because $\Delta = \partial_r^2 + 2r^{-1}\partial_r + r^{-2}\Delta^*$, we have

$$\begin{aligned}
-\gamma^2 H_{n,j}(r) &= \int_{\mathbb{S}^2} Y_{n,j}(\xi)(-\gamma^2 h(r\xi))\,\mathrm{d}\omega(\xi) = \int_{\mathbb{S}^2} Y_{n,j}(\xi)\Delta h(r\xi)\,\mathrm{d}\omega(\xi) \\
&= \int_{\mathbb{S}^2} Y_{n,j}(\xi)\partial_r^2 h(r\xi)\,\mathrm{d}\omega(\xi) + \frac{2}{r}\int_{\mathbb{S}^2} Y_{n,j}(\xi)\partial_r h(r\xi)\,\mathrm{d}\omega(\xi) \\
&\quad + \frac{1}{r^2}\int_{\mathbb{S}^2} Y_{n,j}(\xi)\Delta^* h(r\xi)\,\mathrm{d}\omega(\xi) \\
&= H''_{n,j}(r) + \frac{2}{r}H'_{n,j}(r) + \frac{1}{r^2}\int_{\mathbb{S}^2} h(r\xi)\Delta^* Y_{n,j}(\xi)\,\mathrm{d}\omega(\xi),
\end{aligned}$$

where we have used Lemma 3.3.5 in the latter equation. The fact that $\Delta^* Y_{n,j}(\xi) = -n(n+1)Y_{n,j}(\xi)$ proves the first part of the lemma. For the second part of the lemma we need to prove that a function $f \in \mathrm{C}^2[0,R)$ which solves

$$f''(r) + \frac{2}{r} f'(r) - \frac{n(n+1)}{r^2} f(r) = -\gamma^2 f(r) \quad \text{for } r > 0$$

has the form $f(r) = j_n(\gamma r)$. For this we subsitute $f(r) =: G(\gamma r)$ and $t := \gamma r$. If we multiply the differential equation for f with r^2, we arrive at

$$r^2 f''(r) + 2r f'(r) + (\gamma^2 r^2 - n(n+1)) f(r) = 0.$$

Plugging in t and G instead of r and f yields

$$t^2 G''(t) + 2t G'(t) + (t^2 - n(n+1)) G(t) = 0.$$

This is the well known differential equation for the spherical Bessel functions (see 10.1.1 in [1] on page 437), so the general solution has the form

$$G(t) = c_1 j_n(t) + c_2 y_n(t) \quad \text{for some constants } c_1, c_2 \in \mathbb{R}.$$

See 10.1.1 on page 437 in [1] for a definition of the functions y_n. Since the function f is continuous at 0, so is G. This means that $c_2 = 0$, because the funtions y_n all have a singularity at 0. We conclude that $f(r) = c_1 j_n(\gamma r)$. So our functions $H_{n,j}$ must have the required form. □

Theorem 3.3.7 *If $h \in C^2(B_R(0))$ satisfies the Helmholtz equation with parameter $\gamma > 0$, then it has the form*

$$h(r\xi) = \sum_{n=0}^{\infty} \sum_{j=-n}^{n} c_{n,j} j_n(\gamma r) Y_{n,j}(\xi),$$

where the convergence is absolute and uniform in every closed proper subset of $B_R(0)$.

Proof. We know by Lemma 3.3.6 that the functions

$$H_{n,j}(r) := \int_{\mathbb{S}^2} h(r\xi) Y_{n,j}(\xi) \, d\omega(\xi)$$

satisfy $H_{n,j}(r) = c_{n,j} j_n(\gamma r)$ for some constants $c_{n,j} \in \mathbb{R}$. We then write $Y_n(\xi) := \sum_{j=-n}^{n} c_{n,j} Y_{n,j}(\xi)$ and define numbers $C_n > 0$ by

$$C_n^2 := \int_{\mathbb{S}^2} |Y_n(\xi)|^2 \, \mathrm{d}\omega(\xi) = \sum_{j=-n}^{n} |c_{n,j}|^2,$$

which holds because the $Y_{n,j}$ form an orthonormal basis of $\mathrm{L}^2(\mathbb{S}^2)$. Now for every $r_0 < R$ the function $\xi \mapsto h(r_0\xi)$ is in $\mathrm{L}^2(\mathbb{S}^2)$, so

$$\infty > \int_{\mathbb{S}^2} |h(r_0\xi)|^2 \, \mathrm{d}\omega(\xi) = \sum_{n=0}^{\infty} \sum_{j=-n}^{n} |c_{n,j}|^2 |j_n(\gamma r_0)|^2 = \sum_{n=0}^{\infty} |j_n(\gamma r_0)|^2 C_n^2.$$

Here we have used the fact that $H_{n,j}(r_0) = \langle h(r_0 \cdot), Y_{n,j} \rangle_{\mathrm{L}^2(\mathbb{S}^2)}$. The above equation implies that the series $\sum_{n=0}^{\infty} |j_n(\gamma r_0)|^2 C_n^2$ converges. So we can use Lemma 3.3.4 to conclude that the function

$$\psi(r\xi) := \sum_{n=0}^{\infty} \sum_{j=-n}^{n} c_{n,j} j_n(\gamma r) Y_{n,j}(\xi)$$

converges uniformly in every closed subset of $B_R(0)$. In particular this means it is continuous in ξ for every fixed r. Applying Lemma 2.2.9 together with

$$\langle h(r\cdot) - \psi(r\cdot), Y_{n,j} \rangle_{\mathrm{L}^2(\mathbb{S}^2)} = 0$$

for every n, j and fixed r proves that $h(r\xi) = \psi(r\xi)$. □

4 Modelling the Wind Inside the Planet

If we want to model the wind within a gas giant, the **Euler equations** in a rotating reference frame together with **mass conservation** are the right choice [11, 19].

$$\partial_t \boldsymbol{u} + (\boldsymbol{u} \cdot \nabla)\,\boldsymbol{u} + 2\boldsymbol{\Omega} \times \boldsymbol{u} + \boldsymbol{\Omega} \times (\boldsymbol{\Omega} \times \boldsymbol{x}) = -\frac{1}{F}\nabla p + \boldsymbol{g} \text{ for } \boldsymbol{x} \in B, \tag{4.1}$$

$$\partial_t F + \nabla \cdot (F\boldsymbol{u}) = 0 \text{ for } \boldsymbol{x} \in B. \tag{4.2}$$

The quantities that these equations relate are:

- $\boldsymbol{u} = \boldsymbol{u}(\boldsymbol{x}, t) \in \mathrm{C}^2(\overline{B} \times \mathbb{R}^+, \mathbb{R}^3)$ is the velocity of the wind,
- $\boldsymbol{\Omega} = \Omega \boldsymbol{\varepsilon}^3$ is the constant angular rotation rate of the planet,
- $F = F(\boldsymbol{x}, t) \in \mathrm{C}^2(\overline{B} \times \mathbb{R}^+)$ is the mass density from Chapter 3 satisfying Assumption 3.1.1,
- $p = p(\boldsymbol{x}, t) \in \mathrm{C}^2(\overline{B} \times \mathbb{R}^+)$ is the pressure and
- $\boldsymbol{g} = \boldsymbol{g}(\boldsymbol{x}, t)$ is the gravitational acceleration.

Here $\boldsymbol{x} \in \mathbb{R}^3$ represents location and $t \geq 0$ is the time. B is the interior of a regular surface $\Sigma \subseteq \mathbb{R}^3$. The terms of Equation (4.1) have the following physical interpretations:

- $\partial_t \boldsymbol{u}$ is the acceleration of the fluid (the wind),
- $(\boldsymbol{u} \cdot \nabla)\,\boldsymbol{u}$ corresponds to the inertial acceleration, the first two terms together form the material derivative of the wind velocity,
- The next two terms are caused by fictious forces due to rotation, the first one is the Coriolis acceleration $2\boldsymbol{\Omega} \times \boldsymbol{u}$ and

T. Peter, *Reconstructing Wind Fields from Gravitational Data on Gas Giants*, BestMasters, https://doi.org/10.1007/978-3-658-51277-4_4

- $\boldsymbol{\Omega} \times (\boldsymbol{\Omega} \times \boldsymbol{x})$ is the centrifugal acceleration.
- On the right hand side we find the acceleration due to the pressure gradient $-\frac{1}{F}\nabla p$ and
- the gravitational acceleration $\boldsymbol{g}$ which is caused by the gravitational potential V from Definition 2.3.2, so $\boldsymbol{g} = -\nabla V$.

It is worth noting that we know the wind field on the surface of the planet and will assume that this quantity and its derivatives up to order 2 are continuous on ∂B (See for example [4, 17]). Equation (4.1) is nonlinear in $\boldsymbol{u}$ and thus too hard to solve so we need some further simplifications.

4.1 Simplifying the Euler Equation

Assumption 4.1.1 We assume that Equations (4.1) and (4.2) are **time independent**. In particular, this means that $\partial_t \boldsymbol{u} = 0$ and $\partial_t F = 0$ on $B \times \mathbb{R}^+$. We will omit the time argument in all physical quantities from now on.

Assumption 4.1.2 We assume that the system has a **small Rossby number**. The Rossby number is the ratio between the inertial and the Coriolis acceleration. If it is small, then the Coriolis term in Equation (4.1) is dominant and we can neglect the inertial term, meaning that $(\boldsymbol{u} \cdot \nabla)\boldsymbol{u} \approx 0$.

We can further simplify Equation (4.1) by writing the centrifugal acceleration as a gradient field.

Definition 4.1.3 The **centrifugal potential** V_Ω is defined as

$$V_\Omega(\boldsymbol{x}) := -\frac{1}{2}\Omega^2 \left(x_1^2 + x_2^2\right).$$

Lemma 4.1.4 *The centrifugal acceleration is a potential field, specifically*

$$\nabla V_\Omega(\boldsymbol{x}) = \boldsymbol{\Omega} \times (\boldsymbol{\Omega} \times \boldsymbol{x}).$$

Proof. We simply calculate the right-hand side of the above equation:

$$\begin{aligned}\boldsymbol{\Omega} \times (\boldsymbol{\Omega} \times \boldsymbol{x}) &= \Omega^2 \boldsymbol{\varepsilon}^3 \times (\boldsymbol{\varepsilon}^3 \times (x_1 \boldsymbol{\varepsilon}^1 + x_2 \boldsymbol{\varepsilon}^2 + x_3 \boldsymbol{\varepsilon}^3)) \\ &= \Omega^2 (\boldsymbol{\varepsilon}^3 \times (x_1 \boldsymbol{\varepsilon}^2 - x_2 \boldsymbol{\varepsilon}^1)) \\ &= -\Omega^2 (x_1 \boldsymbol{\varepsilon}^1 + x_2 \boldsymbol{\varepsilon}^2) \\ &= \nabla V_\Omega(\boldsymbol{x}).\end{aligned}$$

This proves the lemma above. □

With these simplifications we can restate Equation (4.1) to make it easier to solve.

Theorem 4.1.5 (unperturbed thermal wind equation) *If Equation* (4.1) *as well as Assumptions 4.1.1 and 4.1.2 are satisfied, then the following equations also (approximately) hold.*

$$2\boldsymbol{\Omega} \times (F\boldsymbol{u}) = -\nabla p - F\nabla(V + V_\Omega) \text{ on } B \text{ and} \tag{4.3}$$

$$2(\boldsymbol{\Omega} \cdot \nabla)(F\boldsymbol{u}) = \nabla F \times \nabla(V + V_\Omega) \text{ on } B. \tag{4.4}$$

Of course mass conservation (4.2) *holds as well.*

Proof. With the above assumptions and Definition 4.1.3, Equation (4.1) becomes

$$\begin{aligned}2\boldsymbol{\Omega} \times (F\boldsymbol{u}) + F\nabla V_\Omega &= -\nabla p - F\nabla V \\ \Leftrightarrow 2\boldsymbol{\Omega} \times (F\boldsymbol{u}) &= -\nabla p - F\nabla(V + V_\Omega).\end{aligned}$$

For the second equation, we take the curl of Equation (4.3). For the right-hand side of course $\nabla \times \nabla p = 0$ and by the product rule
$\nabla \times (F\nabla(V + V_\Omega)) = \nabla F \times \nabla(V + V_\Omega)$, because $\nabla \times \nabla(V + V_\Omega) = 0$. For the left-hand side it is easy to check that

$$\begin{aligned}\nabla \times (\boldsymbol{\Omega} \times (F\boldsymbol{u})) &= \boldsymbol{\Omega}\nabla \cdot (F\boldsymbol{u}) - (\boldsymbol{\Omega} \cdot \nabla)(F\boldsymbol{u}) \\ &= -(\boldsymbol{\Omega} \cdot \nabla)(F\boldsymbol{u}),\end{aligned}$$

where the last equality is due to mass conservation (4.2). Using these two results the curl of Equation (4.3) is

$$2(\boldsymbol{\Omega} \cdot \nabla)(F\boldsymbol{u}) = \nabla F \times \nabla(V + V_{\Omega}),$$

which proves the theorem. □

The partial differential equation from above is still too complicated, because the right-hand side is nonlinear in F (V also depends on F). To have a chance to solve this equation, we need to linearize it around some static solution.

4.2 Perturbation of the Thermal Wind Equation

We now want to simplify further by using a perturbation approach on Equations (4.3) and (4.4).

Definition 4.2.1 We define the background or **static quantities** $F_0, p_0 \in C^2(\overline{B})$ and V_0 (caused by F_0 as in Definition 2.3.2) as those physical quantities which solve Equation (4.3) (and therefore Equation (4.4)) for $\boldsymbol{u} = 0$, meaning

$$\begin{aligned} 0 &= -\nabla p_0 - F_0 \nabla(V_0 + V_{\Omega}) \quad \text{and as a consequence} \\ 0 &= \nabla F_0 \times \nabla(V_0 + V_{\Omega}) \quad \text{are both satisfied on } B. \end{aligned}$$

Definition 4.2.2 We define the **dynamical quantities** or perturbations F', V' and p' as

$$\begin{aligned} F'(\boldsymbol{x}) &:= F(\boldsymbol{x}) - F_0(\boldsymbol{x}) \\ V'(\boldsymbol{x}) &:= V(\boldsymbol{x}) - V_0(\boldsymbol{x}) \\ p'(\boldsymbol{x}) &:= p(\boldsymbol{x}) - p_0(\boldsymbol{x}). \end{aligned}$$

Here F, V and p satisfy Theorem 4.1.5. The interpretation is that F', V' and p' are purely caused by the wind field $\boldsymbol{u}$. Because of the linearity of the gravitational potential, V' and F' satisfy the relation in Definition 2.3.2.

With these definitions we can simplify the equations from the previous chapter.

Theorem 4.2.3 (perturbed thermal wind equation) *If we neglect the second order of the perturbations in Equation (4.4), then we get a linear partial differential equation in the quantities F', V' and $\boldsymbol{u}$. We have*

$$2(\boldsymbol{\Omega}\cdot\nabla)(F_0\boldsymbol{u}) = \nabla F' \times \nabla(V_0 + V_\Omega) - \nabla V' \times \nabla F_0 \text{ on } B. \qquad (4.5)$$

This is the most general form of the perturbed thermal wind equation.

Proof. We plug the expansions $F = F_0 + F'$ and $V = V' + V_0$ into Equation (4.4) and neglect the terms of second order in F', V', p' and $\boldsymbol{u}$.

$$\begin{aligned}
2(\boldsymbol{\Omega}\cdot\nabla)((F_0+F')\boldsymbol{u}) &= \nabla(F_0+F')\times\nabla(V_0+V'+V_\Omega)\\
\Leftrightarrow 2(\boldsymbol{\Omega}\cdot\nabla)(F_0\boldsymbol{u}) + 2(\boldsymbol{\Omega}\cdot\nabla)(F'\boldsymbol{u}) &= \nabla F_0\times\nabla(V_0+V_\Omega) + \nabla F_0\times\nabla V'\\
&\quad + \nabla F'\times\nabla(V_0+V_\Omega) + \nabla F'\times\nabla V'\\
\Leftrightarrow 2(\boldsymbol{\Omega}\cdot\nabla)(F_0\boldsymbol{u}) + 2(\boldsymbol{\Omega}\cdot\nabla)(F'\boldsymbol{u}) &= \nabla F_0\times\nabla V' + \nabla F'\times\nabla(V_0+V_\Omega)\\
&\quad + \nabla F'\times\nabla V'\\
\Rightarrow \qquad 2(\boldsymbol{\Omega}\cdot\nabla)(F_0\boldsymbol{u}) &= \nabla F_0\times\nabla V' + \nabla F'\times\nabla(V_0+V_\Omega)\\
\Leftrightarrow \qquad 2(\boldsymbol{\Omega}\cdot\nabla)(F_0\boldsymbol{u}) &= \nabla F'\times\nabla(V_0+V_\Omega) - \nabla V'\times\nabla F_0.
\end{aligned}$$

We have used $0 = \nabla F_0 \times \nabla(V_0 + V_\Omega)$ from Definition 4.2.1 in the second step and neglected the terms involving $F'\boldsymbol{u}$ and $\nabla F' \times \nabla V'$, which are quadratic in the perturbations, in the third step. □

We have now split up the problem into two smaller problems: First we need to determine the static quantities F_0, V_0 and p_0 from the equations in Definition 4.2.1. For this an appropiate model for a rigidly rotating gas giant is needed. Having determined the static quantities, we can plug these into the equations in Theorem 4.2.3 and try to determine F' from $\boldsymbol{u}$.

The Static Model 5

The static quantities p_0, $F_0 \in C^2(\overline{B})$ and V_0, as defined in the previous section, are the solutions to the **hydrostatic equilibrium** (see Definition 4.2.1)

$$\nabla p_0 = -F_0 \nabla(V_0 + V_\Omega) \text{ on } B. \tag{5.1}$$

The approach we take here considers a simple model for the relationship between pressure and density.

5.1 Polytrope of Index Unity

Assumption 5.1.1 In the following we will assume that the static density and pressure obey a **polytrope of index unity**, which means that there exists some constant $K > 0$ such that

$$p_0 = K F_0^2 \text{ on } B.$$

In general a polytrope of index $n \in \mathbb{N}$ refers to the relation $p_0 = K F_0^{1+1/n}$.

Theorem 5.1.2 *If the static quantities from Definition 4.2.1 (these functions are the solutions to Equation* (5.1)*) satisfy Assumption 5.1.1, then the density satisfies the inhomogeneous Helmholtz equation*

$$\left(\Delta + \frac{2\pi G}{K}\right) F_0 = \frac{\Omega^2}{K} \text{ on } B.$$

T. Peter, *Reconstructing Wind Fields from Gravitational Data on Gas Giants*, BestMasters, https://doi.org/10.1007/978-3-658-51277-4_5

Proof. Assumption 5.1.1 lets us compute the pressure gradient in terms of the density.

$$\nabla p_0 = K \nabla F_0^2 = 2K F_0 \nabla F_0 \text{ on } B.$$

Therefore Equation (5.1) implies that

$$\begin{aligned} 2K F_0 \nabla F_0 &= -F_0 \nabla (V_0 + V_\Omega) \\ \Rightarrow \quad 2K \nabla F_0 &= -\nabla (V_0 + V_\Omega) = -\nabla U \text{ on } B, \end{aligned}$$

where $U(\boldsymbol{x}) := V_0(\boldsymbol{x}) + V_\Omega(\boldsymbol{x})$. We now take the divergence on both sides of the above equation. Because of the properties of the gravitational potential we summarized after Definition 2.3.2, the equality $\Delta V_0 = 4\pi G F_0$ holds on B. It is also easy to check that $\Delta V_\Omega = -2\Omega^2$, such that $\Delta U = 4\pi G F_0 - 2\Omega^2$. Plugging this all into the above equation, we get

$$\begin{aligned} 2K \Delta F_0 &= -\Delta U = -4\pi G F_0 + 2\Omega^2 \\ \Leftrightarrow 2K \Delta F_0 + 4\pi G F_0 &= 2\Omega^2 \\ \Leftrightarrow \quad \left(\Delta + \frac{2\pi G}{K}\right) F_0 &= \frac{\Omega^2}{K} \text{ on } B, \end{aligned}$$

so the specified Helmholtz equation is indeed satisfied. □

Theorem 5.1.3 *Under the same conditions as in Theorem 5.1.2 and if* $B = B_R(0)$, *the static density* F_0 *has the form*

$$F_0(r\boldsymbol{\xi}) = \overline{F} \left(\frac{2q}{3} + \sum_{n=0}^{\infty} b_n j_n \left(\frac{\alpha\pi}{R} r \right) P_n(\xi_3) \right).$$

Here $\overline{F} := 3M/(4\pi R^3)$ *is the mean density of the planet,* $q := \Omega^2 R^3/(GM)$ *and* $\alpha := \sqrt{2GR^2/(\pi K)}$. *The* b_n *form a sequence of real numbers, such that the convergence of the series is uniform in every closed proper subset of* B.

Proof. First we define the function f_0 by

$$f_0 : B \to \mathbb{R}_0^+,\ \boldsymbol{x} \mapsto \frac{F_0(\boldsymbol{x})}{\overline{F}}.$$

Then because $\frac{2\pi G}{K} = \frac{\alpha^2\pi^2}{R^2}$ and $\frac{\Omega^2}{K} = \frac{\overline{F}}{R^2}\alpha^2\pi^2\frac{2q}{3}$ the Helmholtz equation from Theorem 5.1.2 becomes

$$\left(\Delta + \frac{\alpha^2\pi^2}{R^2}\right) f_0(\boldsymbol{x}) = \frac{\alpha^2\pi^2}{R^2}\frac{2q}{3} \quad \text{for every } \boldsymbol{x} \in B.$$

If we define the function $h(\boldsymbol{x}) := f_0(\boldsymbol{x}) - \frac{2q}{3}$, then this function satisfies

$$\left(\Delta + \left(\frac{\alpha\pi}{R}\right)^2\right) h = 0 \quad \text{on } B.$$

Since $h \in \mathrm{C}^2(B)$, we can apply Theorem 3.3.7 to conclude that

$$h(r\boldsymbol{\xi}) = \sum_{n=0}^{\infty}\sum_{j=-n}^{n} c_{n,j}\, j_n\left(\frac{\alpha\pi r}{R}\right) Y_{n,j}(\boldsymbol{\xi}),$$

where the convergence is uniform in closed proper subsets of B. We now fix an arbitrary $m \in \mathbb{N}_0$ and choose a number $r_0 < R$ such that $j_m(\alpha\pi r_0/R) \neq 0$. Then we can calculate the coefficients $c_{m,k}$ by

$$\int_{\mathbb{S}^2} h(r_0\boldsymbol{\xi})Y_{m,k}(\boldsymbol{\xi})\, \mathrm{d}\omega(\boldsymbol{\xi}) = \sum_{n=0}^{\infty}\sum_{j=-n}^{n} c_{n,j}\, j_n\left(\frac{\alpha\pi r_0}{R}\right)\delta_{n,m}\delta_{j,k}$$

$$\Leftrightarrow \frac{\int_{\mathbb{S}^2} h(r_0\boldsymbol{\xi})Y_{m,k}(\boldsymbol{\xi})\, \mathrm{d}\omega(\boldsymbol{\xi})}{j_m\left(\frac{\alpha\pi r_0}{R}\right)} = c_{m,k},$$

where we used the uniform convergence of the series and the orthonormality of the spherical harmonics. Since F_0 does not depend on φ by Assumption 3.1.1, f_0 and h do not depend on φ as well. So integrals of the form

$$\int_{\mathbb{S}^2} h(r\boldsymbol{\xi})Y_{n,j}(\boldsymbol{\xi})\, \mathrm{d}\omega(\boldsymbol{\xi})$$

include a factor

$$\int_0^{2\pi} \sin(j\varphi)\, \mathrm{d}\varphi \quad \text{or} \quad \int_0^{2\pi} \cos(j\varphi)\, \mathrm{d}\varphi,$$

when using spherical coordinates. Both of these integrals vanish if $j \neq 0$. This means that $c_{n,j} = 0$ if $j \neq 0$, so if we define $b_n := \sqrt{\frac{2n+1}{4\pi}} c_{n,0}$, then

$$h(r\boldsymbol{\xi}) = \sum_{n=0}^{\infty} b_n j_n \left(\frac{\alpha \pi r}{R}\right) P_n(\xi_3)$$

$$\Rightarrow F_0(r\boldsymbol{\xi}) = \overline{F} \left(\frac{2q}{3} + \sum_{n=0}^{\infty} b_n j_n \left(\frac{\alpha \pi}{R} r\right) P_n(\xi_3)\right),$$

which proves the theorem. □

We now present a way to calculate the potential gradient if the density is known.

Lemma 5.1.4 *If ρ_0, F_0, V_0 and V_Ω satisfy Equation* (5.1) *and Assumption 5.1.1, then we can calculate the potential gradient by*

$$\nabla(V_0 + V_\Omega) = -2K \nabla F_0 \text{ on } B.$$

Proof. As before we use the fact that $\nabla \rho_0 = 2K F_0 \nabla F_0$ and plug this into Equation (5.1). We arrive at

$$2K F_0 \nabla F_0 = -F_0 \nabla(V_0 + V_\Omega)$$

$$\Rightarrow \quad 2K \nabla F_0 = -\nabla(V_0 + V_\Omega).$$

This proves the desired relation. □

The parameters α and $(b_n)_{n \in \mathbb{N}_0}$, which are necessary to fully determine the functions F_0 and ∇V_0 can be computed using the methods presented in [22, 23]. We now present the simplest possible case, which assumes radial symmetry.

5.2 The Case of Radial Symmetry

In the following we will assume that Assumption 5.1.1 holds. Additionally, we assume

Assumption 5.2.1 The mass density F_0 only depends on $|\boldsymbol{x}|$ or r in spherical coordinates.

This allows us the simplify the result of Theorem 5.1.3.

Theorem 5.2.2 *If the static mass density F_0 satisfies the prerequisites of Theorem 5.1.3 as well as Assumption 5.2.1, then*

$$F_0(\boldsymbol{x}) = F_0(r\boldsymbol{\xi}) = \overline{F}\left(\frac{2q}{3} + b_0 j_0\left(\frac{\alpha\pi}{R}r\right)\right)$$

with the same notation as in Theorem 5.1.3.

Proof. Of course we use Theorem 5.1.3 to start with the expression

$$F_0(r\boldsymbol{\xi}) = \overline{F}\left(\frac{2q}{3} + \sum_{n=0}^{\infty} b_n j_n\left(\frac{\alpha\pi}{R}r\right) P_n(\xi_3)\right).$$

Since the convergence of this series is uniform in balls which are proper subsets of B, we can switch series and integral in the following term which uses spherical coordinates.

$$\begin{aligned}
&\frac{2m+1}{2}\int_{-1}^{1} F_0(r\boldsymbol{\xi}(\varphi,t))P_m(t)\,\mathrm{d}t \\
=&\frac{2m+1}{2}\overline{F}\left(\frac{2q}{3}\int_{-1}^{1} P_m(t)\,\mathrm{d}t + \sum_{n=0}^{\infty} b_n j_n\left(\frac{\alpha\pi}{R}r\right)\int_{-1}^{1} P_n(t)P_m(t)\,\mathrm{d}t\right) \\
=&\overline{F}\left(\frac{2q}{3}\delta_{m,0} + b_m j_m\left(\frac{\alpha\pi}{R}r\right)\right).
\end{aligned}$$

On the other hand, since F_0 only depends on r by Assumption 5.2.1 we can pull it out of the integral and get

$$\frac{2m+1}{2}\int_{-1}^{1} F_0(r\xi(\varphi,t))P_m(t)\,\mathrm{d}t = F_0(r\xi(\varphi,t))\delta_{m,0}.$$

So if $m \neq 0$, we conclude that

$$0 = b_m j_m\left(\frac{\alpha\pi}{R}r\right),$$

where this equation holds for every $r \in (0, R)$. Since the spherical Bessel functions do not vanish identically, we conclude that $b_m = 0$ for every number $m \in \mathbb{N}$. □

Lemma 5.2.3 *If the static mass density F_0 satisfies the prerequisites of Theorem 5.1.3 as well as Assumption 5.2.1, then*

$$r\xi \mapsto \frac{\partial F_0(r\xi)}{r}$$

is continuous and does not vanish at $r = 0$.

Proof. First we calculate the radial derivative of F_0. Since

$$j_0(x) = \frac{\sin(x)}{x} \quad \text{and} \quad j_1(x) = \frac{\sin(x) - x\cos(x)}{x^2},$$

we have $j_0'(x) = -j_1(x)$. With an application of the chain rule we arrive at

$$\partial_r F_0(r\xi) = -\overline{F}b_0\frac{\alpha\pi}{R}j_1\left(\frac{\alpha\pi}{R}r\right).$$

To notice that the quantities F_0, $\partial_r F_0$ and $\partial_r F_0/r$ are defined at $r = 0$ requires knowledge of some basic limits.

$$\sin(x) = x - \frac{x^3}{6} + \mathcal{O}(x^5) \quad \text{and} \quad \cos(x) = 1 - \frac{x^2}{2} + \mathcal{O}(x^4) \quad \text{as } x \to 0,$$

which implies that

$$j_0(x) = 1 + \mathcal{O}(x^2) \quad \text{and} \quad j_1(x) = \frac{x}{3} + \mathcal{O}(x^3) \quad \text{as } x \to 0.$$

This allows us to conclude the proof by noting that $j_1(x)/x \to \frac{1}{3}$ as $x \to 0$. □

Remark 5.2.4 We know that $\partial_r F_0$ vanishes if and only if $j_1((\alpha\pi/R)r)$ vanishes and that the smallest positive zero of j_1, called x_1 in this remark, is greather than $\frac{4}{3}\pi$ (see Table 10.6 on page 467 of [1]). So $\partial_r F_0$ would have a zero if $\alpha > \frac{x_1}{\pi} > \frac{4}{3}$. But as we will see in Remark 5.2.5, $|1-\alpha| < 1/20$ holds for both Jupiter and Saturn, meaning that the quotient $\partial_r F_0/r$ does not vanish on all of $B = B_R(0)$.

In the radially symmetric case we have seen that the static density only depends on two unknown parameters, α and b_0 (see Theorem 5.2.2). To determine these numbers, we impose some conditions on F_0, the first one is simply the formula for the total mass

$$\int_B F_0(\boldsymbol{x})\,\mathrm{d}\boldsymbol{x} = M. \tag{5.2}$$

The second condition is also given in [6, 22] and states that

$$F_0(\boldsymbol{x}) = 0 \quad \text{for every } \boldsymbol{x} \in \partial B. \tag{5.3}$$

Of course in our case $B = B_R(0)$, so we can plug in the expression for F_0 from Theorem 5.2.2 and compute the integral over $B_R(0)$ in Equation (5.2) using spherical coordinates. A bit of algebra gets us the two equations

$$0 = \frac{2q}{3} - 1 + \frac{3b_0}{\alpha^3\pi^3}(\sin(\alpha\pi) - \alpha\pi\cos(\alpha\pi)) \tag{5.4}$$

$$0 = \frac{2q}{3}\pi\alpha + b_0\sin(\alpha\pi). \tag{5.5}$$

We can solve these equations for the solution with the smallest non-zero α.

Remark 5.2.5 The solutions to Equations (5.4) and (5.5) with the smallest non-zero α are

- For Jupiter $\alpha \approx 1.0185$ and $b_0 \approx 3.2748$.
- For Saturn $\alpha \approx 1.0327$ and $b_0 \approx 3.2667$.

These values were calculated by first transforming Equations (5.4) and (5.5) into a single equation which has α as the only unknown. Then the value $q = \Omega^2 R^3/(GM)$ was calculated for the respective planet using the data from [20] and [21]. Here R is the equatorial radius. Plugging in the values of q then allows α to be determined. Finally we calculate b_0 from the value of α.

The simplest Form of the Thermal Wind Equation

6

One model commonly used in the literature (in [10, 11]) makes some additional assumptions. In this chapter we will model the planet as $B = B_R(0)$.

Assumption 6.1 We assume that the static density F_0 is spherically symmetric, meaning it only depends on the radial distance r in spherical coordinates. We further assume that the static quantities F_0, V_0 and p_0 satisfy a polytrope of index unity (see Chap. 5).

Assumption 6.2 In this chapter we also assume that the gravitational acceleration caused by the wind-induced density is neglible. This means that

$$\nabla V' \approx 0 \text{ on } B.$$

Lemma 6.3 *If Assumptions and 6.1 and 6.2 hold, then the wind induced density F' satisfies*

$$2(\boldsymbol{\Omega} \cdot \nabla)(F_0(r\boldsymbol{\xi})u_\varphi(r\boldsymbol{\xi})) = 2K \frac{\partial_r F_0(r\boldsymbol{\xi})}{r} \left(\boldsymbol{\varepsilon}^\varphi \cdot \mathrm{L}^*\right) F'(r\boldsymbol{\xi}) \tag{6.1}$$

for every $r\boldsymbol{\xi} \in B \setminus \{0\}$. Here $u_\varphi(r\boldsymbol{\xi}) := \boldsymbol{\varepsilon}^\varphi \cdot \boldsymbol{u}(r\boldsymbol{\xi})$.

Proof. Using Assumption 6.2, Equation (4.5) becomes

$$2(\boldsymbol{\Omega} \cdot \nabla)(F_0\boldsymbol{u}) = \nabla F' \times \nabla(V_0 + V_\Omega) \text{ on } B.$$

Applying Lemma 5.1.4, which is valid because of Assumption 6.1, the equation is transformed into

T. Peter, *Reconstructing Wind Fields from Gravitational Data on Gas Giants*, BestMasters, https://doi.org/10.1007/978-3-658-51277-4_6

$$2(\boldsymbol{\Omega} \cdot \nabla)(F_0 \boldsymbol{u}) = 2K \nabla F_0 \times \nabla F' \text{ on } B.$$

Now we use Assumption 6.1 to simplify the gradient of F_0. $\nabla F_0 = \boldsymbol{\xi} \partial_r F_0$ and because $\boldsymbol{\xi} \times \nabla = \frac{1}{r} \mathrm{L}^*$ we get the relation

$$2(\boldsymbol{\Omega} \cdot \nabla)(F_0(r\boldsymbol{\xi}) \boldsymbol{u}(r\boldsymbol{\xi})) = 2K \frac{\partial_r F_0(r\boldsymbol{\xi})}{r} \mathrm{L}^* F'(r\boldsymbol{\xi}).$$

We note that because of Lemma 2.1.9 the operator L^* only has a component in the direction of $\boldsymbol{\varepsilon}^\varphi$. That is why we will take the inner product with $\boldsymbol{\varepsilon}^\varphi$ on both sides, yielding

$$\left[2(\boldsymbol{\Omega} \cdot \nabla)(F_0(r\boldsymbol{\xi}) \boldsymbol{u}(r\boldsymbol{\xi}))\right] \cdot \boldsymbol{\varepsilon}^\varphi = 2K \frac{\partial_r F_0(r\boldsymbol{\xi})}{r} \left(\boldsymbol{\varepsilon}^\varphi \cdot \mathrm{L}^*\right) F'(r\boldsymbol{\xi}).$$

Since $\boldsymbol{\Omega} = \Omega \boldsymbol{\varepsilon}^3$ we get $\boldsymbol{\Omega} \cdot \nabla = \Omega \partial_3$. Combined with $\partial_3 \boldsymbol{\varepsilon}^\varphi = 0$ this means that we can pull the inner product into the differential operator and get the desired result. □

To explicitly calculate the wind-induced density F' we define a related quantity that will also be used in the next chapter.

Definition 6.4 We define the **thermal-wind source term** G^u as the function

$$G^u(r\boldsymbol{\xi}(\varphi, t)) := -\frac{4\pi G}{K} \frac{r}{\partial_r F_0(r\boldsymbol{\xi}(\varphi, t))} \int_{-1}^{t} \frac{(\boldsymbol{\Omega} \cdot \nabla)(F_0 u_\varphi)(r\boldsymbol{\xi}(\varphi, \tau))}{\sqrt{1 - \tau^2}} \, \mathrm{d}\tau,$$

where $r \in [0, R]$, $\varphi \in [0, 2\pi)$ and $t \in [-1, 1]$ are spherical coordinates.

Remark 6.5 The function G^u from Definition 6.4 is well defined on $B = B_R(0)$ in the case of a radially symmetric polytropic model, as presented in Sect. 5.2. This fact is based on two observations. The first one involves the quotient

$$r \mapsto \frac{r}{\partial_r F_0(r\boldsymbol{\xi})},$$

which only depends on r since we assume the static model is radially symmetric. We have proven that this quotient is well-defined at $r = 0$ in Lemma 5.2.3. In Remark 5.2.4 we also saw that $\partial_r F_0$ has no zeros on B. So in this case the above quotient is well-defined on all of B. The second observation involves the integral

$$\int_{-1}^{t} \frac{(\boldsymbol{\Omega} \cdot \nabla)(F_0 u_\varphi)(r\boldsymbol{\xi}(\varphi, \tau))}{\sqrt{1-\tau^2}} \, d\tau.$$

We assumed that both F_0 and $\boldsymbol{u}$ are twice continuously differentiable on $\overline{B}$. This means that the function

$$r\boldsymbol{\xi} \mapsto (\boldsymbol{\Omega} \cdot \nabla)(F_0(r\boldsymbol{\xi}) u_\varphi(r\boldsymbol{\xi}))$$

is bounded on $\overline{B}$. In particular, the function

$$t \mapsto (\boldsymbol{\Omega} \cdot \nabla)(F_0(r\boldsymbol{\xi}(\varphi, t)) u_\varphi(r\boldsymbol{\xi}(\varphi, t)))$$

is then bounded on $[-1, 1]$ for every $r \in [0, R]$ (it is independent of φ by Assumption 3.1.1). If we combine this with the fact that the function

$$\tau \mapsto (1-\tau^2)^{-1/2}$$

is integrable on $[-1, 1]$, then we conclude that

$$\tau \mapsto \frac{(\boldsymbol{\Omega} \cdot \nabla)(F_0 u_\varphi)(r\boldsymbol{\xi}(\varphi, \tau))}{\sqrt{1-\tau^2}}$$

is integrable over $[-1, 1]$ for every $r \in [0, R]$ and $\varphi \in [0, 2\pi)$ in spherical coordinates. This proves that G^u on the whole is well-defined on $\overline{B}$. We can also conclude that it is continuous on this set.

Theorem 6.6 *Under Assumptions 6.1 and 6.2 we can calculate the latitudinal derivative of F' by*

$$\partial_t F'(r\boldsymbol{\xi}(\varphi, t)) = -\frac{r}{K \partial_r F_0(r\boldsymbol{\xi}(\varphi, t))} \frac{(\boldsymbol{\Omega} \cdot \nabla)(F_0 u_\varphi)(r\boldsymbol{\xi}(\varphi, t))}{\sqrt{1-t^2}}.$$

Here $0 \le r < R$, $\varphi \in [0, 2\pi)$ and $t \in (-1, 1)$ (although all the quantities are of course independent of φ by Assumption 3.1.1). This means we can determine the wind-induced density F' up to an additive function which only depends on r – called $\eta(r)$ – by the following formula.

$$F'(r\boldsymbol{\xi}(\varphi, t)) = \frac{G^u(r\boldsymbol{\xi}(\varphi, t))}{4\pi G} + \eta(r)$$

Proof. For the first formula we use Lemma 6.3 and divide by $2K\partial_r F_0/r$. We can do this because of Lemma 5.2.3 and Remark 5.2.4. So we arrive at

$$\frac{r}{K\partial_r F_0(r\xi(\varphi,t))}(\boldsymbol{\Omega}\cdot\nabla)(F_0 u_\varphi)(r\xi(\varphi,t)) = (\boldsymbol{\varepsilon}^\varphi \cdot \mathrm{L}^*)F'(r\xi(\varphi,t)).$$

By Lemma 2.1.9, we know that $\mathrm{L}^* F' = \boldsymbol{\varepsilon}^\varphi(\boldsymbol{\varepsilon}^\varphi \cdot \mathrm{L}^*)F'$ and therefore

$$\boldsymbol{\varepsilon}^\varphi \cdot \mathrm{L}^* F'(r\xi(\varphi,t)) = -\sqrt{1-t^2}\,\partial_t F'(r\xi(\varphi,t)).$$

Now we just need to plug this into our equation and divide by $-\sqrt{1-t^2}$, which is possible for $t \in (-1,1)$. This gets us the first equation we wanted to derive. For the second equation, we just need to integrate the first one with respect to the latitude t and notice that the coefficient $r/(K\partial_r F_0)$ does not depend on t. This means we can pull this factor out of the integral and are then left with the quantity $G^u/(4\pi G)$. Of course we also get a constant of integration which will depend on the remaining coordinates r and φ. Since we assume φ-independence, this constant will only depend on r and we name it as $\eta(r)$. □

The Thermo-Gravitational Wind Equation 7

Another model which has been considered in the literature (by [19]) is the **thermo-gravitational wind equation**. It is based on deriving a Helmholtz equation for the wind-induced gravitational potential V'. This model also presumes $B = B_R(0)$.

7.1 Derivation of the Helmholtz Equation

We first present the necessary assumptions we need to make this model work.

Assumption 7.1.1 We assume that the static density F_0 is spherically symmetric, meaning it only depends on the radial distance r in spherical coordinates. We further assume that the static quantities F_0, V_0 and p_0 satisfy a polytrope of index unity (see Chap. 5).

The main difference in this modelling to the previous one is that we do not neglect the contribution of the wind-induced gravitational potential V' to the force balance.

Lemma 7.1.2 *If Assumption 7.1.1 holds, then the wind-induced density F' and the corresponding gravitational potential V' satisfy*

$$2(\boldsymbol{\Omega} \cdot \nabla)(F_0(r)u_\varphi(r\boldsymbol{\xi})) = \frac{\partial_r F_0(r\boldsymbol{\xi})}{r} \left(\boldsymbol{\varepsilon}^\varphi \cdot \mathrm{L}^*\right) (2KF'(r\boldsymbol{\xi}) + V'(r\boldsymbol{\xi})) \tag{7.1}$$

for every $r\boldsymbol{\xi} \in B$.

T. Peter, *Reconstructing Wind Fields from Gravitational Data on Gas Giants*, BestMasters, https://doi.org/10.1007/978-3-658-51277-4_7

Proof. We start out with the perturbed thermal wind equation (4.5) and substitute $-2K\nabla F_0$ for $\nabla(V_0 + V_\Omega)$ (Lemma 5.1.4), just as in the last section.

$$\begin{aligned} &2(\boldsymbol{\Omega}\cdot\nabla)(F_0(r\boldsymbol{\xi})\boldsymbol{u}(r\boldsymbol{\xi})) = 2K\nabla F_0(r\boldsymbol{\xi}) \times \nabla F'(r\boldsymbol{\xi}) + \nabla F_0(r\boldsymbol{\xi}) \times \nabla V'(r\boldsymbol{\xi}) \\ \Leftrightarrow\ &2(\boldsymbol{\Omega}\cdot\nabla)(F_0(r\boldsymbol{\xi})\boldsymbol{u}(r\boldsymbol{\xi})) = \nabla F_0(r\boldsymbol{\xi}) \times \nabla(2KF'(r\boldsymbol{\xi}) + V'(r\boldsymbol{\xi})). \end{aligned}$$

Using the fact that F_0 only depends on r, we derive $\nabla F_0 = \boldsymbol{\xi}\partial_r F_0$ and therefore

$$2(\boldsymbol{\Omega}\cdot\nabla)(F_0(r\boldsymbol{\xi})\boldsymbol{u}(r\boldsymbol{\xi})) = \frac{\partial_r F_0(r)}{r}\mathrm{L}^*(2KF'(r\boldsymbol{\xi}) + V'(r\boldsymbol{\xi})).$$

Again because of Lemma 2.1.9 we take the inner product with $\boldsymbol{\varepsilon}^\varphi$ on both sides and arrive at the statement we wanted to prove, exactly as in Lemma 6.3. □

Lemma 7.1.3 *The wind-induced potential satifies the Helmholtz equation*

$$\Delta V'(r\boldsymbol{\xi}) + \frac{2\pi G}{K}V'(r\boldsymbol{\xi}) = G^u(r\boldsymbol{\xi}) + \eta(r) \quad \textit{for all} \quad r\boldsymbol{\xi} \in B_R(0). \tag{7.2}$$

Here $\eta(r)$ is an undetermined function which only depends on r. For every $r\boldsymbol{\xi} \in \mathbb{R}^3 \setminus \overline{B_R(0)}$ the wind induced potential satisfies the Laplace equation

$$\Delta V'(r\boldsymbol{\xi}) = 0.$$

Proof. The wind-induced potential V' is caused by the wind-induced density F' as in Definition 2.3.2, meaning that $\Delta V' = 4\pi GF'$ in $B_R(0)$ and $\Delta V' = 0$ on $\mathbb{R}^3 \setminus \overline{B_R(0)}$. This proves the second equation of the lemma. For the first equation, we assume $r\boldsymbol{\xi} \in B_R(0)$, then the derivation is very similar to the proof of Theorem 6.6. We start with Equation (7.1) and notice that $(\boldsymbol{\varepsilon}^\varphi \cdot \mathrm{L}^*)F' = -\sqrt{1-t^2}\partial_t F'$ by Lemma 2.1.9. So we can manipulate the equation in exactly the same way as in the proof of Theorem 6.6 arriving at

$$-\frac{2r}{\partial_r F_0(r\boldsymbol{\xi}(\varphi,t))}\frac{(\boldsymbol{\Omega}\cdot\nabla)(F_0 u_\varphi)(r\boldsymbol{\xi}(\varphi,t))}{\sqrt{1-t^2}} = \partial_t(2KF'(r\boldsymbol{\xi}(\varphi,t)) + V'(r\boldsymbol{\xi}(\varphi,t)))$$

for every $0 \le r < R$, $0 \le \varphi < 2\pi$ and $-1 < t < 1$. Again we integrate both sides with respect to t, then the coefficient $-2r/\partial_r F_0$ on the left-hand side does not depend on the integration variable. Also we get a constant of integration c which only depends on r. With Definition 6.4 this results in

$$\frac{K}{2\pi G} G^u(r\xi(\varphi,t)) + c(r) = 2KF'(r\xi(\varphi,t)) + V'(r\xi(\varphi,t)).$$

Now since $\Delta V' = 4\pi GF'$ is satisfied on $B_R(0)$ we get

$$\begin{aligned} &\frac{K}{2\pi G}\Delta V'(r\xi) + V'(r\xi) = \frac{K}{2\pi G} G^u(r\xi) + c(r) \\ \Leftrightarrow \quad &\left(\Delta + \frac{2\pi G}{K}\right) V'(r\xi) = G^u(r\xi) + \eta(r), \end{aligned}$$

where $\eta(r) := 2\pi Gc(r)/K$. This proves the lemma. □

To proceed further we need to develop the theory of solutions to an inhomogeneous Helmholtz equation as in Equation (7.2).

7.2 An Orthonormal Basis for the Gravitational Potential

In this chapter B will refer to the set $B_R(0) \subseteq \mathbb{R}^3$ for some radius $R > 0$. The problem we want to solve can be stated as follows.

Definition 7.2.1 For a given source term $f \in \mathrm{C}^2(B)$ and constant $\beta \geq 0$ we want to solve an equation of the form

$$\Delta V(\boldsymbol{x}) + \beta^2 V(\boldsymbol{x}) = f(\boldsymbol{x}) \quad \text{for every } \boldsymbol{x} \in B,$$

where the solution $V \in \mathrm{C}^2(B) \cap \mathrm{C}^1(\mathbb{R}^3) \cap \mathrm{C}^2(\mathbb{R}^3 \setminus \overline{B})$ is a gravitational potential, which implies V is regular at infinity and solves $\Delta V = 0$ in $\mathbb{R}^3 \setminus \overline{B}$.

We will solve this equation by developing an orthonormal basis of $\mathrm{L}^2(B)$ which is in some sense "similar" to the gravitational potential. By this we mean that our basis functions should fufill some boundary conditions that are motivated by the gravitational potential. As we will show now the gravitational potential is actually fully determined on the outside of B once we know the value on the boundary ∂B.

Theorem 7.2.2 *The gravitational potential $V \in \mathrm{C}^2(\mathbb{R}^3 \setminus \overline{B})$ which satisfies*

$$\Delta V(\boldsymbol{x}) = 0 \quad \textit{for every } \boldsymbol{x} \in \mathbb{R}^3 \setminus \overline{B}$$

and is regular at infinity is determined by the formula

$$V(r\boldsymbol{\xi}) = \sum_{n=0}^{\infty} \sum_{j=-n}^{n} \langle V(R\,\cdot), Y_{n,j} \rangle_{\mathrm{L}^2(\mathbb{S}^2)} \left(\frac{R}{r}\right)^{n+1} Y_{n,j}(\boldsymbol{\xi})$$

for every $r > R$.

Proof. The proof of this is easily adapted from the explanations in [14], Sect. 3.4 (see page 136 for the formula). □

To further proceed, we derive some boundary conditions for V on ∂B, which will motivate the boundary conditions for our basis, as previously stated.

Lemma 7.2.3 *If the gravitational potential V satisfies all the conditions of Theorem 7.2.2 as well as $V \in \mathrm{C}^1(\mathbb{R}^3)$, then in the sense of $\mathrm{L}^2(\mathbb{S}^2)$ we have*

$$V(R\boldsymbol{\xi}) = \sum_{n=0}^{\infty} \sum_{j=-n}^{n} \langle V(R\,\cdot), Y_{n,j} \rangle_{\mathrm{L}^2(\mathbb{S}^2)} Y_{n,j}(\boldsymbol{\xi}) \quad \textit{and}$$

$$\partial_r V(r\boldsymbol{\xi})\big|_{r=R} = -\sum_{n=0}^{\infty} \sum_{j=-n}^{n} \frac{n+1}{R} \langle V(R\,\cdot), Y_{n,j} \rangle_{\mathrm{L}^2(\mathbb{S}^2)} Y_{n,j}(\boldsymbol{\xi}).$$

Proof. The first equation follows directly because $V \in \mathrm{C}^1(\mathbb{R}^3) \Rightarrow V(R\,\cdot) \in \mathrm{C}^1(\mathbb{S}^2) \subseteq \mathrm{L}^2(\mathbb{S}^2)$. For the second equation we need to prove first that we can differentiate the series termwise with respect to r. This is possible if the series over the differentiated terms converges uniformly and the series itself converges at least at one point (see Theorem 104.6 on page 554 of [5]). To prove this we take a fixed, but arbitrary $r_0 > R$ and note first that for every $r > r_0$ and any large $N \in \mathbb{N}$ we have

$$\begin{aligned}
&\left| \sum_{n=N+1}^{\infty} \sum_{j=-n}^{n} \langle V(R\,\cdot), Y_{n,j} \rangle_{\mathrm{L}^2(\mathbb{S}^2)} \left(\frac{R}{r}\right)^{n+1} Y_{n,j}(\boldsymbol{\xi}) \right| \\
&\leq \sum_{n=N+1}^{\infty} \sum_{j=-n}^{n} \left| \langle V(R\,\cdot), Y_{n,j} \rangle_{\mathrm{L}^2(\mathbb{S}^2)} \right| \left|\frac{R}{r}\right|^{n+1} |Y_{n,j}(\boldsymbol{\xi})| \\
&\leq \sum_{n=N+1}^{\infty} \sum_{j=-n}^{n} \|V(R\,\cdot)\|_{\mathrm{L}^2(\mathbb{S}^2)} \left|\frac{R}{r}\right|^{n+1} \sqrt{\frac{2n+1}{4\pi}} \\
&\leq \frac{\|V(R\,\cdot)\|_{\mathrm{L}^2(\mathbb{S}^2)}}{\sqrt{4\pi}} \sum_{n=N+1}^{\infty} (2n+1)^{3/2} \left|\frac{R}{r_0}\right|^{n+1} \to 0 \quad \text{for } N \to \infty,
\end{aligned}$$

where the convergence to 0 is independent of r and $\boldsymbol{\xi}$. In the second inequality we used the Cauchy-Schwarz inequality and Lemma 2.2.8. This shows the uniform convergence of the series representing V on $\mathbb{R}^3 \setminus \overline{B_{r_0}(0)}$, so this series converges pointwise on this set. The argument for the differentiated series is almost the same, because for $r > r_0$ and large $N \in \mathbb{N}$ we have

$$\begin{aligned}
&\left| \sum_{n=N+1}^{\infty} \sum_{j=-n}^{n} \langle V(R\,\cdot), Y_{n,j} \rangle_{\mathrm{L}^2(\mathbb{S}^2)} \partial_r \left(\frac{R}{r}\right)^{n+1} Y_{n,j}(\boldsymbol{\xi}) \right| \\
&= \left| \sum_{n=N+1}^{\infty} \sum_{j=-n}^{n} \frac{n+1}{R} \langle V(R\,\cdot), Y_{n,j} \rangle_{\mathrm{L}^2(\mathbb{S}^2)} \left(\frac{R}{r}\right)^{n+2} Y_{n,j}(\boldsymbol{\xi}) \right| \\
&\leq \sum_{n=N+1}^{\infty} \sum_{j=-n}^{n} \frac{n+1}{R} \|V(R\,\cdot)\|_{\mathrm{L}^2(\mathbb{S}^2)} \left|\frac{R}{r}\right|^{n+2} \sqrt{\frac{2n+1}{4\pi}} \\
&\leq \frac{\|V(R\,\cdot)\|_{\mathrm{L}^2(\mathbb{S}^2)}}{\sqrt{4\pi}\,R^2} \sum_{n=N+1}^{\infty} (2n+1)^{5/2} \left|\frac{R}{r_0}\right|^{n+2} \to 0 \quad \text{for } N \to \infty.
\end{aligned}$$

Again the convergence is independent of r and $\boldsymbol{\xi}$, so this series converges uniformly for every $\boldsymbol{x} \in \mathbb{R}^3$ with $|\boldsymbol{x}| > r_0$. This proves that we can differentiate every term in the series seperately, implying

$$\begin{aligned}\partial_r V(r\boldsymbol{\xi}) &= \sum_{n=0}^{\infty}\sum_{j=-n}^{n} \langle V(R\,\cdot), Y_{n,j}\rangle_{\mathrm{L}^2(\mathbb{S}^2)} \partial_r \left(\frac{R}{r}\right)^{n+1} Y_{n,j}(\boldsymbol{\xi}) \\ &= -\sum_{n=0}^{\infty}\sum_{j=-n}^{n} \frac{n+1}{R} \langle V(R\,\cdot), Y_{n,j}\rangle_{\mathrm{L}^2(\mathbb{S}^2)} \left(\frac{R}{r}\right)^{n+2} Y_{n,j}(\boldsymbol{\xi}).\end{aligned}$$

Since $\partial_r V(r\,\cdot) \in \mathrm{C}(\mathbb{S}^2) \subseteq \mathrm{L}^2(\mathbb{S}^2)$ for every $r \geq R$, we conclude that

$$\langle \partial_r V(r\,\cdot), Y_{n,j}\rangle_{\mathrm{L}^2(\mathbb{S}^2)} = -\frac{n+1}{R}\left(\frac{R}{r}\right)^{n+2} \langle V(R\,\cdot), Y_{n,j}\rangle_{\mathrm{L}^2(\mathbb{S}^2)}$$

for every $n \in \mathbb{N}_0$ and $j \in \{-n, \ldots, n\}$. Because the function

$$r \mapsto \langle \partial_r V(r\,\cdot), Y_{n,j}\rangle_{\mathrm{L}^2(\mathbb{S}^2)}$$

is continuous on $[R, \infty)$ for every valid index n and j, we can take the limit $r \to R$ in the above equation to get

$$\langle \partial_r V(r\,\cdot), Y_{n,j}\rangle_{\mathrm{L}^2(\mathbb{S}^2)}\big|_{r=R} = -\frac{n+1}{R}\langle V(R\,\cdot), Y_{n,j}\rangle_{\mathrm{L}^2(\mathbb{S}^2)}.$$

This proves the lemma above. □

The lemma we just proved will give some motivation for the orthonormal basis we are about to construct. Because of the symmetry of the underlying space we can decompose the inner product of $\mathrm{L}^2(B)$. If $u, v \in \mathrm{L}^2(B)$, we have

$$\langle u, v\rangle_{\mathrm{L}^2(B)} = \int_B u(\boldsymbol{x})v(\boldsymbol{x})\,\mathrm{d}\boldsymbol{x} = \int_0^R r^2 \int_{\mathbb{S}^2} u(r\boldsymbol{\xi})v(r\boldsymbol{\xi})\,\mathrm{d}\omega(\boldsymbol{\xi})\,\mathrm{d}r.$$

This suggests that we should decompose our orthonormal basis similarly, meaning any basis function u should have the form

$$u(\boldsymbol{x}) = u(r\boldsymbol{\xi}) = F(r)Y_{n,j}(\boldsymbol{\xi}),$$

where the $Y_{n,j}$ are the fully normalized spherical harmonics, which form an orthonormal basis of $\mathrm{L}^2(\mathbb{S}^2)$. Two basis functions $u \neq v$ with $u(r\boldsymbol{\xi}) = F(r)Y_{n,j}(\boldsymbol{\xi})$ and $v(r\boldsymbol{\xi}) = G(r)Y_{n,j}(\boldsymbol{\xi})$ must then satisfy

$$\int_0^R r^2 F(r)G(r)\,\mathrm{d}r = 0 \quad \text{and}$$

$$\int_0^R r^2 F^2(r)\,\mathrm{d}r = 1.$$

This means we have to construct an orthonormal basis of the weighted L^2-space $\mathrm{L}^2_w(0, R)$, where the basis functions are twice continuously differentiable and the weight is $w(r) := r^2$. Since we want some similarity to the gravitational potential, we demand a boundary condition motivated by Lemma 7.2.3:
The function F which forms the radial part of one of our basis functions u satisfies $F \in \mathrm{C}^2[0, R]$ and

$$F'(R) = -\frac{n+1}{R} F(R). \tag{7.3}$$

Since we want to develop a gravitational potential into this basis and we often encounter Laplacians when dealing with potentials, we also demand that our basis functions are eigenfunctions of the Laplacian on B. This means any basis function u should satisfy

$$\Delta u = -\lambda u \quad \text{on } B. \tag{7.4}$$

At first we only demand $\lambda \in \mathbb{C}$ but we will restrict this later.

Lemma 7.2.4 *If $u \in \mathrm{C}^2(B \setminus \{0\})$ is of the form $u(\boldsymbol{x}) = u(r\boldsymbol{\xi}) = F(r)Y_{n,j}(\boldsymbol{\xi})$ for some $n \in \mathbb{N}_0$ and $j \in \{-n, \dots, n\}$, then the following statements are equivalent:*

1. $\Delta u(r\boldsymbol{\xi}) = -\lambda u(r\boldsymbol{\xi})$ *for every* $r\boldsymbol{\xi} \in B$.
2. $\mathrm{L}F(r) = -\lambda w(r)F(r)$ *for every* $r \in (0, R)$, *where* $w(r) := r^2$ *and*

$$\mathrm{L}F(r) := \frac{\mathrm{d}}{\mathrm{d}r}\left(r^2 \frac{\mathrm{d}}{\mathrm{d}r} F(r)\right) - n(n+1)F(r).$$

Proof. Since $\Delta = \partial_r^2 + \frac{2}{r}\partial_r + \frac{1}{r^2}\Delta^*$ and $\Delta^* Y_{n,j}(\boldsymbol{\xi}) = -n(n+1)Y_{n,j}(\boldsymbol{\xi})$ for every valid index n and j and finally $w(r) = r^2 > 0$ on $(0, R)$, the following equivalences hold for every $r \in (0, R)$ and $\boldsymbol{\xi} \in \mathbb{S}^2$

$$
\begin{aligned}
&& \mathrm{L}F(r) &= -\lambda w(r)F(r) \\
&\Leftrightarrow && \left(\frac{\mathrm{d}}{\mathrm{d}r}\left(r^2\frac{\mathrm{d}}{\mathrm{d}r}F(r)\right) - n(n+1)F(r)\right)Y_{n,j}(\xi) = -\lambda r^2 F(r)Y_{n,j}(\xi) \\
&\Leftrightarrow && \left(r^2F''(r) + 2rF'(r) - n(n+1)F(r)\right)Y_{n,j}(\xi) = -\lambda r^2 F(r)Y_{n,j}(\xi) \\
&\Leftrightarrow && \left(F''(r) + \frac{2}{r}F'(r) - \frac{n(n+1)}{r^2}F(r)\right)Y_{n,j}(\xi) = -\lambda F(r)Y_{n,j}(\xi) \\
&\Leftrightarrow && \Delta u(r\xi) = -\lambda u(r\xi),
\end{aligned}
$$

where the multiplication by the $Y_{n,j}$ preserves the equivalence since they are not identically zero. □

We are now ready to define the eigenvalue problem which the radial parts F of our basis are supposed to satisfy.

Definition 7.2.5 For any of our basis functions $u(r\xi) = F(r)Y_{n,j}(\xi)$, where $n \in \mathbb{N}_0$ and $j \in \{-n, \dots, n\}$, the functions $F \in \mathrm{C}^2[0, R]$ must satisfy the **eigenvalue problem** $\mathrm{L}F(r) = -\lambda w(r)F(r)$ for every $r \in (0, R)$ and some $\lambda \in \mathbb{C}$. Here

$$\mathrm{L}F(r) := \frac{\mathrm{d}}{\mathrm{d}r}\left(r^2\frac{\mathrm{d}}{\mathrm{d}r}F(r)\right) - n(n+1)F(r), \tag{7.5}$$

with the weight function $w(r) := r^2$. We also impose the boundary conditions

$$\delta_{n,0}F'(0) + (1-\delta_{n,0})F(0) = 0 \quad \text{and} \tag{7.6}$$

$$F'(R) + \frac{n+1}{R}F(R) = 0. \tag{7.7}$$

Here the second boundary condition is motivated by our examination of the gravitational potential and the first one is imposed to provide unique solvability. We will call λ the eigenvalue of the problem and F the eigenfunction. Of course we get a different boundary value problem for every $n \in \mathbb{N}_0$, but we will suppress this in the notation for now.

Now we need to find every eigenvalue and -function of the problem defined in Definition 7.2.5. First we will prove some general properties of the eigenvalues λ.

Theorem 7.2.6 *If the function* $F \in \mathrm{C}^2[0, R]$ *satisfies the eigenvalue problem stated in Definition 7.2.5 for some* $n \in \mathbb{N}_0$*, then the eigenvalue* λ *is a real number and we can choose the eigenfunctions* F *as real functions.*

Proof. First we define the vector space $V \subseteq \mathrm{C}^2[0, R]$ as

$$V := \{h \in \mathrm{C}^2[0, R] \mid h \text{ satisfies (7.6) and (7.7)}\}.$$

Since the boundary conditions are linear with respect to the function we apply them to, this is indeed a vector space. For $F, G \in V$ we define the inner product

$$\langle F, G\rangle_V := \int_0^R F(r)\overline{G(r)}\,\mathrm{d}r.$$

Then we can use integration by parts twice to show that the differential operator L is self-adjoint in $(V, \langle\cdot,\cdot\rangle_V)$.

$$\begin{aligned}
\langle \mathrm{L}F, G\rangle_V &= \int_0^R \frac{\mathrm{d}}{\mathrm{d}r}\left(r^2\frac{\mathrm{d}}{\mathrm{d}r}F(r)\right)\overline{G(r)}\,\mathrm{d}r - n(n+1)\langle F, G\rangle_V \\
&= \left[r^2F'(r)\overline{G(r)}\right]_0^R - \int_0^R F'(r)r^2\overline{G'(r)}\,\mathrm{d}r - n(n+1)\langle F, G\rangle_V \\
&= \left[r^2\left(F'(r)\overline{G(r)} - F(r)\overline{G'(r)}\right)\right]_0^R + \langle F, \mathrm{L}G\rangle_V \\
&= R^2\left(F'(R)\overline{G(R)} - F(R)\overline{G'(R)}\right) + \langle F, \mathrm{L}G\rangle_V.
\end{aligned}$$

Since both F and G are in V, the boundary term satisfies

$$\begin{aligned}
&R^2\left(F'(R)\overline{G(R)} - F(R)\overline{G'(R)}\right) \\
&= - R(n+1)F(R)\overline{G(R)} + R(n+1)F(R)\overline{G(R)} = 0,
\end{aligned}$$

which proves that the operator L is self-adjoint on V. If F is a solution of the eigenvalue problem stated in Definition 7.2.5, we have $F \in V$ and $\mathrm{L}F = -\lambda w F$ on $(0, R)$. Then

$$\begin{aligned}
&-\lambda\langle wF, F\rangle_V = \langle \mathrm{L}F, F\rangle_V = \langle F, \mathrm{L}F\rangle_V = -\overline{\lambda}\langle F, wF\rangle_V \\
\Leftrightarrow \quad &\left(\lambda - \overline{\lambda}\right)\langle wF, F\rangle_V = 0.
\end{aligned}$$

Here the equivalence holds because w is a real function. Since we do not allow F to be identically zero and $w > 0$ on $(0, R)$, the term $\langle wF, F\rangle_V$ is positive meaning that $\lambda = \overline{\lambda}$, so $\lambda \in \mathbb{R}$. Since L is a real differential operator, if F is a (potentially complex) solution to $\mathrm{L}F = -\lambda w F$, so is $\mathrm{Re}(F)$, meaning we can choose real eigenfunctions. □

Remark 7.2.7 Using the notation of the last theorem, we have seen that for $F, G \in V$

$$\langle \mathrm{L}F, G\rangle_V = \langle F, \mathrm{L}G\rangle_V$$

holds. This means that for two eigenfunctions $\mathrm{L}F = -\lambda w F$ and $\mathrm{L} = -\mu w G$, where $\lambda \neq \mu$, we can conclude that

$$-\lambda\langle wF, G\rangle_V = \langle \mathrm{L}F, G\rangle_V = \langle F, \mathrm{L}G\rangle_V = -\mu\langle F, wG\rangle_V = -\mu\langle wF, G\rangle_V.$$

We used the fact that w and the eigenvalues only take real values. Rearranging this equation we get

$$(\lambda - \mu)\langle wF, G\rangle_V = 0.$$

Because of $(\lambda - \mu) \neq 0$, the inner product must be zero. Since the eigenfunctions can also be chosen as real functions such that

$$0 = \langle wF, G\rangle_V = \int_0^R r^2 F(r)G(r)\,\mathrm{d}r = \langle F, G\rangle_{\mathrm{L}^2_w(0,R)},$$

the eigenfunctions for different eigenvalues are orthogonal in $\mathrm{L}^2_w(0, R)$.

Theorem 7.2.8 *If the function $F \in \mathrm{C}^2[0, R]$ satisfies the eigenvalue problem stated in Definition 7.2.5 for some $n \in \mathbb{N}_0$, then the eigenvalue λ is non-negative.*

Proof. We again use the vector space V from the previous theorem

$$V := \{h \in \mathrm{C}^2[0, R] \mid h \text{ satisfies (7.6) and (7.7)}\}$$

and almost the same inner product on this space

$$\langle F, G\rangle_V := \int_0^R F(r)G(r)\,\mathrm{d}r.$$

We have neglected the complex conjugate in the inner product since we already know that all occuring quantities are real. Integrating by parts once and using $F \in V$ we get

$$\begin{aligned}\langle \mathrm{L}F, F\rangle_V &= \int_0^R \frac{\mathrm{d}}{\mathrm{d}r}\left(r^2\frac{\mathrm{d}}{\mathrm{d}r}F(r)\right)F(r)\,\mathrm{d}r - n(n+1)\langle F, F\rangle_V \\ &= \left[r^2F'(r)F(r)\right]_0^R - \int_0^R r^2\left(F'(r)\right)^2\,\mathrm{d}r - n(n+1)\langle F, F\rangle_V \\ &= R^2F'(R)F(R) - \langle wF', F'\rangle_V - n(n+1)\langle F, F\rangle_V.\end{aligned}$$

Since F is in V, it satisfies Equation (7.7). This implies that

$$R^2F'(R)F(R) = -R^2\frac{n+1}{R}F^2(R) = -(n+1)RF^2(R) \leq 0.$$

If we define the function $q(r) := r$, then $q^2 = w$ and

$$\langle \mathrm{L}F, F\rangle_V = -\left((n+1)RF^2(R) + \|qF'\|_V^2 + n(n+1)\|F\|_V^2\right) \leq 0.$$

Now if $\mathrm{L}F = -\lambda wF$ for any $F \in V$ and $\lambda \in \mathbb{R}$, then

$$\begin{aligned}-\lambda\|qF\|_V^2 &= \langle \mathrm{L}F, F\rangle_V \leq 0 \\ &\Rightarrow \lambda \geq 0,\end{aligned}$$

which shows that the eigenvalues are indeed non-negative. □

In the following we will use the notation $\lambda = \gamma^2$, where $\gamma \geq 0$ is some real number, for the eigenvalue λ from Definition 7.2.5. We are now ready to solve the eigenvalue problem presented in Definition 7.2.5.

Theorem 7.2.9 *The solutions $F \in \mathrm{C}^2[0, R]$ to the eigenvalue problem for in Definition 7.2.5 are*

$$F(r) := F_{m,n}(r) := c_{m,n}\,j_n(\gamma_{n,m}r) \quad \textit{for } n \in \mathbb{N}_0 \textit{ and } m \in \mathbb{N},$$

where the $c_{m,n}$ are some real constants and the $\gamma_{n,m}$ are the positive solutions of

$$j_{n-1}(\gamma_{n,m}R) = 0 \quad \text{for } n \in \mathbb{N}_0 \text{ and } m \in \mathbb{N}.$$

The eigenvalues λ corresponding to $F = F_{m,n}$ are $\lambda = \gamma_{n,m}^2$ and these values are distinct for every $n \in \mathbb{N}_0$ and $m \in \mathbb{N}$.

Proof. We first consider the case $\gamma = 0$ for an arbitrary $n \in \mathbb{N}_0$. In this case the differential equation for F becomes

$$r^2 F''(r) + 2r F'(r) - n(n+1)F(r) = 0 \quad \text{for every } r \in (0, R).$$

This is an Euler differential equation, which can be solved with the substitution $F(r) =: G(\ln(r))$. Then our solution is

$$F(r) = c_1 r^n + c_2 r^{-n-1}.$$

Now applying boundary condition (7.6) yields $c_2 = 0$, since for every $n \in \mathbb{N}_0$ our solution has to be continuously differentiable at $r = 0$. The second boundary condition (7.7) can be simplified to

$$c_1(2n+1)R^{n-1} = 0,$$

which can only be true if $c_1 = 0$, so we conclude that $F(r) = 0$ for every $r \in (0, R)$. This means the case $\gamma = 0$ only admits the trivial solution $F = 0$ for every $n \in \mathbb{N}_0$ and will not be considered further. Now for the more interesting case $\gamma > 0$, for an arbitrary $n \in \mathbb{N}_0$. Our differential equation in this case is

$$r^2 F''(r) + 2r F'(r) - n(n+1)F(r) + \gamma^2 r^2 F(r) = 0,$$

which can be transformed into the differential equation for the spherical Bessel functions of order n by the transformation $F(r) =: G(\gamma r)$ and $s := \gamma r$. Our general solution is

$$F(r) = c_1 j_n(\gamma r) + c_2 y_n(\gamma r)$$

(see 10.1.1 in [1] for example). Boundary condition (7.6) is either $F'(0) = 0$ in the case $n = 0$ or $F(0) = 0$ in the case $n \in \mathbb{N}$. The functions j_n satisfy these conditions automatically, but the functions y_n are singular at $r = 0$, so we conclude that $c_2 = 0$.

For the second boundary condition we use the recurrence relation found in 10.1.21 in [1]

$$j_{n-1}(x) = j_n'(x) + \frac{n+1}{x} j_n(x) \quad \text{for every } n \in \mathbb{N}_0 \text{ and } x > 0.$$

Condition (7.7) is thus equivalent to

$$\begin{aligned} F'(R) + \frac{n+1}{R} F(R) &= 0 \\ \Leftrightarrow \quad c_1 \gamma j_n'(\gamma R) + c_1 \frac{n+1}{R} j_n(\gamma R) &= 0 \\ \Leftrightarrow c_1 \gamma \left(j_n'(\gamma R) + \frac{n+1}{\gamma R} j_n(\gamma R) \right) &= 0 \\ \Leftrightarrow \qquad c_1 \gamma j_{n-1}(\gamma R) &= 0 \\ \Leftrightarrow \qquad c_1 j_{n-1}(\gamma R) &= 0. \end{aligned}$$

Since the trivial solution $F = 0$ is uninteresting, we conclude that $c_1 \neq 0$ and therefore $\gamma > 0$ must satisfy $j_{n-1}(\gamma R) = 0$. This equation is satisfied if and only if $J_{n-1/2}(\gamma R) = 0$, which has a countable number of ascending solutions $\gamma_{n,m}$. These solutions are distinct for every $n, m \in \mathbb{N}$ (this follows from 9.5.2 from [1] as well as 15.21, 15.22 and the fifth paragraph in 15.28 from [18]). □

Lemma 7.2.10 *The solutions $F_{m,n}$ from Theorem 7.2.9 satisfy*

$$\| F_{m,n} \|^2_{\mathrm{L}^2_w(0,R)} = \int_0^R r^2 j_n^2(\gamma_{n,m} r)\, \mathrm{d}r = \frac{R^3 j_n^2(\gamma_{n,m} R)}{2},$$

if $c_{m,n} = 1$. This implies that the functions

$$r \mapsto \sqrt{\frac{2}{R^3 j_n^2(\gamma_{n,m} R)}}\, j_n(\gamma_{n,m} r)$$

are normalized in $\mathrm{L}^2_w(0, R)$ for every $n \in \mathbb{N}_0$ and $m \in \mathbb{N}$.

Proof. If we define $\lambda_{n,m} := \gamma_{n,m} R$, then these coefficients satisfy $j_{n-1}(\lambda_{n,m}) = 0$ for every $m \in \mathbb{N}$ and $n \in \mathbb{N}_0$. Therefore using the substitution $s := r/R$ implies

$$\int_0^R r^2 j_n^2(\gamma_{n,m} r)\,\mathrm{d}r = R^3 \int_0^1 s^2 j_n^2(\lambda_{n,m} s)\,\mathrm{d}s,$$

meaning we need to calculate the coefficients

$$d_{n,m} := R^3 \int_0^1 s^2 j_n^2(\lambda_{n,m} s)\,\mathrm{d}s, \quad \text{where}$$
$$0 = j_{n-1}(\lambda_{n,m}) \quad \text{for every } n \in \mathbb{N}_0 \text{ and } m \in \mathbb{N}.$$

Now since $j_n(x) = \sqrt{\frac{\pi}{2x}} J_{n+1/2}(x)$

$$d_{n,m} = \frac{\pi R^3}{2\lambda_{n,m}} \int_0^1 s J_{n+1/2}^2(\lambda_{n,m} s)\,\mathrm{d}s, \quad \text{where}$$
$$0 = J_{n-1/2}(\lambda_{n,m}) \quad \text{for every } n \in \mathbb{N}_0 \text{ and } m \in \mathbb{N}.$$

Using Lemma B.1, we conclude that the $\lambda_{n,m}$ are the positive solutions to the equation

$$J'_{n+1/2}(x) + \frac{n+1/2}{x} J_{n+1/2}(x) = 0 \Leftrightarrow \left(n + \frac{1}{2}\right) J_{n+1/2}(x) + x J'_{n+1/2}(x) = 0,$$

which means we can apply Theorem B.2 to conclude that

$$d_{n,m} = \frac{\pi R^3}{4\lambda_{n,m}} J_{n+1/2}^2(\lambda_{n,m}) = \frac{R^3}{2} j_n^2(\lambda_{n,m}) = \frac{R^3}{2} j_n^2(\gamma_{n,m} R),$$

proving the lemma. □

Theorem 7.2.11 *For a fixed but arbitrary $n \in \mathbb{N}_0$ and with the definitions from Theorem 7.2.9, the sequence of functions*

$$G_{m,n}(r) := \sqrt{\frac{2}{R^3 j_n^2(\gamma_{n,m} R)}}\, j_n(\gamma_{n,m} r),$$

where $m \in \mathbb{N}$, forms an orthonormal basis of $\mathrm{L}^2_w(0, R)$.

Proof. We have seen in Theorem 7.2.9 that the functions $G_{m,n}$ are eigenfunctions of the Problem 7.2.5 for eigenvalues $\gamma_{n,m}^2$, which are all distinct. Therefore we apply

Remark 7.2.7 to conclude that $(G_{m,n})_{m\in\mathbb{N}}$ is an orthogonal sequence in $\mathrm{L}^2_w(0, R)$. This sequence is also orthonormal by Lemma 7.2.10. What remains to be proven is the completeness in $\mathrm{L}^2_w(0, R)$. To do this, we first take an arbitrary $F \in \mathrm{C}^2_c(0, R) \subseteq \mathrm{L}^2_w(0, R)$ (functions which are twice continuously differentiable and have compact support in $(0, R)$) and show that the partial Fourier sum

$$F_M(r) := \sum_{m=1}^{M} \langle F, G_{m,n}\rangle_{\mathrm{L}^2_w(0,R)} G_{m,n}(r)$$

converges to F in $\mathrm{L}^2_w(0, R)$ as $M \to \infty$. We want to use Theorem B.3, so we need to transform from spherical to ordinary Bessel functions and from the interval $(0, R)$ to $(0, 1)$. For the second transformation we substitute $s := r/R$ and $\lambda_{n,m} := \gamma_{n,m}R$. We have seen in the proof of Lemma 7.2.10 that these numbers are the positive solutions of

$$\left(n + \frac{1}{2}\right) J_{n+1/2}(x) + x J'_{n+1/2}(x) = 0.$$

We also define

$$\begin{aligned} f(s) &:= F(Rs) \quad \text{and} \\ f_M(s) &:= \sum_{m=1}^{M} \langle F, G_{m,n}\rangle_{\mathrm{L}^2_w(0,R)} G_{m,n}(Rs) \end{aligned}$$

for every $s \in [0, 1]$ and $M \in \mathbb{N}$. Then we have

$$G_{m,n}(Rs) = \sqrt{\frac{2}{R^3 j_n^2(\lambda_{n,m})}}\, j_n(\lambda_{n,m}s) = \frac{1}{\sqrt{s}}\sqrt{\frac{2}{R^3 J^2_{n+1/2}(\lambda_{n,m})}}\, J_{n+1/2}(\lambda_{n,m}s).$$

With the substitution $s = r/R$ we can also transform the Fourier coefficients.

$$\begin{aligned} \langle F, G_{m,n}\rangle_{\mathrm{L}^2_w(0,R)} &= R^3 \int_0^1 s^2 F(Rs) G_{m,n}(Rs)\,\mathrm{d}s \\ &= R^{3/2}\sqrt{\frac{2}{J^2_{n+1/2}(\lambda_{n,m})}} \int_0^1 s\left(\sqrt{s} f(s)\right) J_{n+1/2}(\lambda_{n,m}s)\,\mathrm{d}s. \end{aligned}$$

For ease of notation we define

$$c_m := \int_0^1 s\left(\sqrt{s} f(s)\right) J_{n+1/2}(\lambda_{n,m} s)\,\mathrm{d}s \quad \text{for every } m \in \mathbb{N}.$$

Plugging this all into the expression for $f_M(s)$ yields

$$\sqrt{s} f_M(s) = \sum_{m=1}^{M} \frac{2}{J_{n+1/2}^2(\lambda_{n,m})} c_m J_{n+1/2}(\lambda_{n,m} s),$$

which looks like the Dini-Bessel series from Theorem B.3. It is easy to see that the function f is in $\mathrm{C}_{\mathrm{c}}^2(0, 1)$. If we define $g(s) := \sqrt{s} f(s)$, then this function vanishes in a neighborhood of $s = 0$, since $\operatorname{supp}(g) = \operatorname{supp}(f) \subsetneq (0, 1)$. We conclude that $g \in \mathrm{C}^1[0, 1]$, so we can apply Theorem B.3, which proves that $\sqrt{s} f_M(s) \to \sqrt{s} f(s)$ as $M \to \infty$ for every $s \in (0, 1)$. It is then obvious that $F_M(r) \to F(r)$ as $M \to \infty$ for every $r \in (0, R)$. We will now show that the sequence F_M is also convergent in $\mathrm{L}_w^2(0, R)$, which means that the pointwise and the $\mathrm{L}_w^2(0, R)$-limit must be equal. We can quantify the decay of the Fourier coefficients using Theorem B.7, because $f \in \mathrm{C}^1[0, 1]$ and

$$\begin{aligned}\langle F, G_{m,n}\rangle_{\mathrm{L}_w^2(0,R)} &= R^3 \int_0^1 s^2 F(Rs) G_{m,n}(Rs)\,\mathrm{d}s \\ &= R^{3/2} \sqrt{\frac{2}{j_n^2(\lambda_{n,m})}} \int_0^1 s^2 f(s) j_n(\lambda_{n,m} s)\,\mathrm{d}s,\end{aligned}$$

where we used the substitution $s = r/R$ again. Therefore a natural number K and a constant $C > 0$ exist, such that

$$\left|\langle F, G_{m,n}\rangle_{\mathrm{L}_w^2(0,R)}\right| \le \frac{C}{m - 3/4} \quad \text{for every } m \ge K.$$

Now, for two natural numbers $N > M \ge K$, we use the orthonormality of the $G_{m,n}$ to conclude that

$$\|F_N - F_M\|_{\mathrm{L}_w^2(0,R)}^2 = \left\| \sum_{m=M+1}^{N} \langle F, G_{m,n}\rangle_{\mathrm{L}_w^2(0,R)} G_{m,n} \right\|_{\mathrm{L}_w^2(0,R)}^2$$

$$= \sum_{m=M+1}^{N} \left|\langle F, G_{m,n}\rangle_{\mathrm{L}_w^2(0,R)}\right|^2 \le \sum_{m=M+1}^{N} \frac{C^2}{(m - 3/4)^2} \xrightarrow[N,M\to\infty]{} 0.$$

We have proven that $(F_M)_{M\in\mathbb{N}}$ is a Cauchy sequence in $\mathrm{L}^2_w(0,R)$, meaning the limit

$$\sum_{m=1}^{\infty}\langle F, G_{m,n}\rangle_{\mathrm{L}^2_w(0,R)} G_{m,n}$$

exists as a $\mathrm{L}^2_w(0,R)$-function. It is well known that there will be a subsequence which converges pointwise almost everywhere to this $\mathrm{L}^2_w(0,R)$-limit. But this subsequence will also converge to F pointwise, as we have previously seen. So we conclude that F is indeed the $\mathrm{L}^2_w(0,R)$-limit of the sequence $(F_M)_{M\in\mathbb{N}}$. If we now take an arbitrary $F\in\mathrm{L}^2_w(0,R)$ and $\varepsilon>0$, then since $\mathrm{C}^2_\mathrm{c}(0,R)$ is dense in $\mathrm{L}^2_w(0,R)$, we can find a $G\in\mathrm{C}^2_\mathrm{c}(0,R)$ such that $\|F-G\|_{\mathrm{L}^2_w(0,R)}\le\varepsilon/3$. Let F_M and G_M represent the M-th partial Fourier sum of F and G, respectively. Then we can choose a natural number $N\in\mathbb{N}$ such that $\|G-G_M\|_{\mathrm{L}^2_w(0,R)}\le\varepsilon/3$ for every $M\ge N$, since $G\in\mathrm{C}^2_\mathrm{c}(0,R)$. This means that

$$\begin{aligned}\|F-F_M\|_{\mathrm{L}^2_w(0,R)} &\le \|F-G\|_{\mathrm{L}^2_w(0,R)}+\|G-G_M\|_{\mathrm{L}^2_w(0,R)}+\|G_M-F_M\|_{\mathrm{L}^2_w(0,R)}\\ &\le\frac{2\varepsilon}{3}+\|F-G\|_{\mathrm{L}^2_w(0,R)}\le\varepsilon\end{aligned}$$

for every $M\ge N$. We used Bessel's inequality in the last line to estimate

$$\|G_M-F_M\|^2_{\mathrm{L}^2_w(0,R)}=\sum_{m=1}^{M}\left|\langle G-F, G_{m,n}\rangle_{\mathrm{L}^2_w(0,R)}\right|^2\le\|F-G\|^2_{\mathrm{L}^2_w(0,R)}$$

This shows the completeness of the orthonormal set $(G_{m,n})_{m\in\mathbb{N}}$. □

Theorem 7.2.12 *With the definitions from Theorem 7.2.11, the functions*

$$u_{m,n,j}(\boldsymbol{x}):=u_{m,n,j}(r\boldsymbol{\xi}):=G_{m,n}(r)Y_{n,j}(\boldsymbol{\xi})$$

for $m\in\mathbb{N}$, $n\in\mathbb{N}_0$ and $j\in\{-n,\ldots,n\}$ form an orthonormal basis of $\mathrm{L}^2(B)$. These functions satisfy

$$\Delta u_{m,n,j}(r\boldsymbol{\xi})=-\gamma^2_{n,m}u_{m,n,j}(r\boldsymbol{\xi})$$

for every valid index m, n, j and $r\boldsymbol{\xi}\in B$.

Proof. The eigenvalue property for the Laplacian is clear from Definition 7.2.5 and Lemma 7.2.4. The normalization of these functions is given because

$$\int_B u_{m,n,j}^2(x)\,\mathrm{d}x = \int_0^R r^2 G_{n,m}^2(r)\,\mathrm{d}r \int_{\mathbb{S}^2} Y_{n,j}^2(\xi)\,\mathrm{d}\omega(\xi) = 1.$$

The orthogonality is derived similarly using the orthogonality of the $Y_{n,j}$ and the $G_{m,n}$ in their respective L^2-spaces. If $(m, n, j) \neq (k, l, i)$

$$\langle u_{m,n,j}, u_{k,l,i}\rangle_{\mathrm{L}^2(B)} = \int_0^R r^2 G_{m,n}(r) G_{k,l}(r)\,\mathrm{d}r \int_{\mathbb{S}^2} Y_{n,j}(\xi) Y_{l,i}(\xi)\,\mathrm{d}\omega(\xi).$$

Now if $n \neq l$ or $j \neq i$, the integral over $\mathbb{S}^2$ vanishes and if $n = l$ and $j = i$, then $m \neq k$ must hold. This implies that the integral over the radial functions must vanish, which proves the orthogonality of the $u_{m,n,j}$. Now for the completeness of the $\mathrm{L}^2(B)$-orthonormal set $(u_{m,n,j})_{m,n,j}$, we start with an arbitrary function $V \in \mathrm{L}^2(B)$ which satisfies $\langle V, u_{m,n,j}\rangle_{\mathrm{L}^2(B)} = 0$ for every $m \in \mathbb{N}, n \in \mathbb{N}_0$ and $j \in \{-n, \dots, n\}$. Because of Fubini's theorem and

$$\infty > \|V\|_{\mathrm{L}^2(B)}^2 = \int_0^R \int_{\mathbb{S}^2} r^2 V^2(r\xi)\,\mathrm{d}\omega(\xi)\,\mathrm{d}r$$

the function $\xi \mapsto rV(r\xi)$ is in $\mathrm{L}^2(\mathbb{S}^2)$ for almost every $r \in (0, R)$, meaning that $\xi \mapsto V(r\xi)$ is also in $\mathrm{L}^2(\mathbb{S}^2)$ for almost every $r \in (0, R)$. We can then expand it into the fully normalized spherical harmonics

$$V(r\,\cdot) = \sum_{n=0}^{\infty} \sum_{j=-n}^{n} \langle V(r\,\cdot), Y_{n,j}\rangle_{\mathrm{L}^2(\mathbb{S}^2)} Y_{n,j}.$$

Defining $V_{n,j}(r) := \langle V(r\,\cdot), Y_{n,j}\rangle_{\mathrm{L}^2(\mathbb{S}^2)}$ for every $n \in \mathbb{N}_0$ and $j \in \{-n, \dots, n\}$, an application of the Cauchy-Schwarz inequality and the fact that the $Y_{n,j}$ are normalized shows that

$$\int_0^R r^2 V_{n,j}^2(r)\,\mathrm{d}r \le \int_0^R r^2 \|V(r\,\cdot)\|_{\mathrm{L}^2(\mathbb{S}^2)}^2\,\mathrm{d}r = \int_B V^2(x)\,\mathrm{d}x = \|V\|_{\mathrm{L}^2(B)}^2 < \infty.$$

Therefore we can expand $V_{n,j}$ into the basis functions $G_{m,n}$, so

$$V_{n,j} = \sum_{m=1}^{\infty} \langle V_{n,j}, G_{m,n} \rangle_{\mathrm{L}^2_w(0,R)} G_{m,n}.$$

The coefficients in this series have the form

$$\begin{aligned}
\langle V_{n,j}, G_{m,n} \rangle_{\mathrm{L}^2_w(0,R)} &= \int_0^R r^2 V_{n,j}(r) G_{m,n}(r)\, \mathrm{d}r \\
&= \int_0^R r^2 \int_{\mathbb{S}^2} G_{m,n}(r) Y_{n,j}(\boldsymbol{\xi}) V(r\boldsymbol{\xi})\, \mathrm{d}\omega(\boldsymbol{\xi})\, \mathrm{d}r \\
&= \int_B u_{m,n,j}(\boldsymbol{x}) V(\boldsymbol{x})\, \mathrm{d}\boldsymbol{x} = \langle V, u_{m,n,j} \rangle_{\mathrm{L}^2(B)} = 0
\end{aligned}$$

for every valid index m, n and j. Thus every single one of the functions $V_{n,j}$ is equal to 0 almost everywhere. Since these functions are also the coefficients in the expansion of $V(r\,\cdot)$ into spherical harmonics we conclude that $V(r\boldsymbol{\xi}) = 0$ for almost every $r \in (0, R)$ and $\boldsymbol{\xi} \in \mathbb{S}^2$. This shows that $V = 0$ in $\mathrm{L}^2(B)$, so the functions $u_{m,n,j}$ indeed form a complete system in this space. □

Finally we present a lemma which illustrates the use of our basis for expanding a gravitational potential.

Lemma 7.2.13 *If $V \in \mathrm{C}^2(B)$ is a gravitational potential as in Definition 2.3.2, then $\Delta V \in \mathrm{L}^2(B)$ and we have*

$$\langle \Delta V, u_{m,n,j} \rangle_{\mathrm{L}^2(B)} = -\gamma_{n,m}^2 \langle V, u_{m,n,j} \rangle_{\mathrm{L}^2(B)}$$

for every $m \in \mathbb{N}$, $n \in \mathbb{N}_0$ and $j \in \{-n, \ldots, n\}$.

Proof. We have $\Delta V = 4\pi G F$, where $F \in \mathrm{C}^2(\overline{B}) \subseteq \mathrm{L}^2(B)$. This proves that $\Delta V \in \mathrm{L}^2(B)$, so the inner products we want to calculate do exist. Now we examine the integral

$$\begin{aligned}
&\langle \Delta V, u_{m,n,j} \rangle_{\mathrm{L}^2(B)} - \langle V, \Delta u_{,m,n,j} \rangle_{\mathrm{L}^2(B)} \\
&= \int_B u_{m,n,j}(\boldsymbol{x}) \Delta V(\boldsymbol{x}) - V(\boldsymbol{x}) \Delta u_{m,n,j}(\boldsymbol{x})\, \mathrm{d}\boldsymbol{x}.
\end{aligned}$$

We want to apply Green's second identity, but since we defined the basis functions in terms of the decomposition $\boldsymbol{x} = r\boldsymbol{\xi}$, which is only valid for $\boldsymbol{x} \neq 0$, we can not include the point $0 \in B$ in our integration domain. We will now use an argument very similar to that made in the proof of Lemma 3.3.5. This means we define the set $D_\varepsilon := B \setminus B_\varepsilon(0)$ for every $\varepsilon > 0$ (see Fig. 3.1), then $u_{m,n,j}$, $V \in \mathrm{C}^2(D_\varepsilon) \cap \mathrm{C}^1(\overline{D_\varepsilon})$, so we can use Green's second identity to conclude that

$$\int_{D_\varepsilon} u_{m,n,j}(\boldsymbol{x})\Delta V(\boldsymbol{x}) - V(\boldsymbol{x})\Delta u_{m,n,j}(\boldsymbol{x})\,\mathrm{d}\boldsymbol{x}$$
$$= \int_{\partial D_\varepsilon} u_{m,n,j}(\boldsymbol{x})\frac{\partial V}{\partial \boldsymbol{n}}(\boldsymbol{x}) - V(\boldsymbol{x})\frac{\partial u_{m,n,j}}{\partial \boldsymbol{n}}(\boldsymbol{x})\,\mathrm{dS}(\boldsymbol{x})$$

for every $\varepsilon > 0$. We will refer to the volume integral by $L_{m,n,j}(\varepsilon)$ and to the surface integral by $R_{m,n,j}(\varepsilon)$. Then by the definition of D_ε we have

$$\left|\int_B u_{m,n,j}(\boldsymbol{x})\Delta V(\boldsymbol{x}) - V(\boldsymbol{x})\Delta u_{m,n,j}(\boldsymbol{x})\,\mathrm{d}\boldsymbol{x} - L_{m,n,j}(\varepsilon)\right|$$
$$\leq \int_{B_\varepsilon(0)} \left|u_{m,n,j}(\boldsymbol{x})\Delta V(\boldsymbol{x}) - V(\boldsymbol{x})\Delta u_{m,n,j}(\boldsymbol{x})\right|\,\mathrm{d}\boldsymbol{x}$$
$$= \int_{B_\varepsilon(0)} \left|4\pi G u_{m,n,j}(\boldsymbol{x})F(\boldsymbol{x}) + \gamma_{n,m}^2 u_{m,n,j}(\boldsymbol{x})V(\boldsymbol{x})\right|\,\mathrm{d}\boldsymbol{x}.$$

Since both V and F are continuous on $\overline{B}$, these functions are bounded. If $\boldsymbol{x} \neq 0$, then $|u_{m,n,j}(\boldsymbol{x})| = |u_{m,n,j}(r\boldsymbol{\xi})| \leq \sqrt{\frac{2n+1}{4\pi}}\|G_{m,n}\|_{\mathrm{C}[0,R]} =: C_{n,m}$ by Lemma 2.2.8. We conclude that the integrand from above is bounded by $C_{n,m}(4\pi G\|F\|_{\mathrm{C}(\overline{B})} + \gamma_{n,m}^2\|V\|_{\mathrm{C}(\overline{B})})$, meaning that the integral as a whole has $K\varepsilon^3$ as an upper bound, where $K > 0$ is some constant. This shows that

$$L_{m,n,j}(\varepsilon) \xrightarrow[\varepsilon\to 0]{} \int_B u_{m,n,j}(\boldsymbol{x})\Delta V(\boldsymbol{x}) - V(\boldsymbol{x})\Delta u_{m,n,j}(\boldsymbol{x})\,\mathrm{d}\boldsymbol{x}.$$

We proceed similarly with $R_{m,n,j}(\varepsilon)$. Since ∂D_ε is the disjoint union of $\partial B_R(0) = \partial B$ and $\partial B_\varepsilon(0)$, we can estimate

$$\left|\int_{\partial B} u_{m,n,j}(\boldsymbol{x})\frac{\partial V}{\partial \boldsymbol{n}}(\boldsymbol{x}) - V(\boldsymbol{x})\frac{\partial u_{m,n,j}}{\partial \boldsymbol{n}}(\boldsymbol{x})\,\mathrm{dS}(\boldsymbol{x}) - R_{m,n,j}(\varepsilon)\right|$$
$$\leq \int_{\partial B_\varepsilon(0)} \left|u_{m,n,j}(\boldsymbol{x})\frac{\partial V}{\partial \boldsymbol{n}}(\boldsymbol{x}) - V(\boldsymbol{x})\frac{\partial u_{m,n,j}}{\partial \boldsymbol{n}}(\boldsymbol{x})\right|\,\mathrm{dS}(\boldsymbol{x}).$$

On $\partial B_\varepsilon(0)$, the outer normal is given as $\boldsymbol{n} = -\boldsymbol{\xi} \in \mathbb{S}^2$, so the normal derivative is $-\partial_r$. Also $x = \varepsilon\boldsymbol{\xi}$ for every $\boldsymbol{x} \in \partial B_\varepsilon(0)$. This implies the integrand from above has an upper bound independent of the integration variable.

$$\begin{aligned}&\left| u_{m,n,j}(\boldsymbol{x})\frac{\partial V}{\partial \boldsymbol{n}}(\boldsymbol{x}) - V(\boldsymbol{x})\frac{\partial u_{m,n,j}}{\partial \boldsymbol{n}}(\boldsymbol{x}) \right| \\ &\leq C_{n,m} \| \, |\nabla V| \, \|_{\mathrm{C}(\overline{B})} + \|V\|_{\mathrm{C}(\overline{B})} D_{n,m},\end{aligned}$$

where $D_{n,m} := \sqrt{\frac{2n+1}{4\pi}} \|G'_{m,n}\|_{\mathrm{C}[0,R]}$ (Here Lemma 2.2.8 was used again). The whole integral can then be estimated by $D\varepsilon^2$ for some constant $D > 0$, which proves that

$$R_{m,n,j}(\varepsilon) \xrightarrow[\varepsilon \to 0]{} \int_{\partial B} u_{m,n,j}(\boldsymbol{x})\frac{\partial V}{\partial \boldsymbol{n}}(\boldsymbol{x}) - V(\boldsymbol{x})\frac{\partial u_{m,n,j}}{\partial \boldsymbol{n}}(\boldsymbol{x}) \, \mathrm{dS}(\boldsymbol{x}).$$

Green's second identity thus also holds on B, if we take the limit $\varepsilon \to 0$, so

$$\begin{aligned}&\langle \Delta V, u_{m,n,j}\rangle_{\mathrm{L}^2(B)} - \langle V, \Delta u_{m,n,j}\rangle_{\mathrm{L}^2(B)} \\ &= \int_{\partial B} u_{m,n,j}(\boldsymbol{x})\frac{\partial V}{\partial \boldsymbol{n}}(\boldsymbol{x}) - V(\boldsymbol{x})\frac{\partial u_{m,n,j}}{\partial \boldsymbol{n}}(\boldsymbol{x}) \, \mathrm{d}\omega(\boldsymbol{x}).\end{aligned}$$

The first boundary integral is calculated by

$$\begin{aligned}&R^2 \int_{\mathbb{S}^2} u_{m,n,j}(R\boldsymbol{\xi})\frac{\partial V}{\partial \boldsymbol{n}}(R\boldsymbol{\xi}) \, \mathrm{d}\omega(\boldsymbol{\xi}) \\ &= R^2 G_{m,n}(R) \int_{\mathbb{S}^2} Y_{n,j}(\boldsymbol{\xi}) \left(\partial_r V(r\boldsymbol{\xi})\right)\big|_{r=R} \, \mathrm{d}\omega(\boldsymbol{\xi}) \\ &= - R^2 \frac{n+1}{R} G_{m,n}(R) \langle V(R\,\cdot), Y_{n,j}\rangle_{\mathrm{L}^2(\mathbb{S}^2)},\end{aligned}$$

where we used Lemma 7.2.3 in the last equality. The other boundary integral can be calculated similarly, so

$$\begin{aligned}&R^2 \int_{\mathbb{S}^2} V(R\boldsymbol{\xi})\frac{\partial u_{m,n,j}}{\partial \boldsymbol{n}}(R\boldsymbol{\xi}) \, \mathrm{d}\omega(\boldsymbol{\xi}) \\ &= R^2 \int_{\mathbb{S}^2} V(R\boldsymbol{\xi}) \left(\partial_r u_{m,n,j}(r\boldsymbol{\xi})\right)\big|_{r=R} \, \mathrm{d}\omega(\boldsymbol{\xi}) \\ &= R^2 G'_{m,n}(R) \langle V(R\,\cdot), Y_{n,j}\rangle_{\mathrm{L}^2(\mathbb{S}^2)}.\end{aligned}$$

Now since our radial functions $G_{m,n}$ satisfy boundary condition (7.7), we conclude that these two terms are actually equal, implying that

$$\langle \Delta V, u_{m,n,j} \rangle_{\mathrm{L}^2(B)} = \langle V, \Delta u_{m,n,j} \rangle_{\mathrm{L}^2(B)} = -\gamma_{n,m}^2 \langle V, u_{m,n,j} \rangle_{\mathrm{L}^2(B)},$$

by Theorem 7.2.12, which proves the statement. □

7.3 Determining the Wind-Induced Density

We now return to the determination of the wind-induced potential V' on B. Of course the quantity we really want to get is F', but once we know V' we can calculate F' via

$$\Delta V'(\boldsymbol{x}) = 4\pi G F'(\boldsymbol{x}) \quad \text{for every } \boldsymbol{x} \in B.$$

As we have seen in Lemma 7.1.3, the wind-induced potential satisfies Equation (7.2) in B. We solve it by expanding both sides into the basis of $\mathrm{L}^2(B)$ from Chap. 7.2.

Lemma 7.3.1 *Equation* (7.2) *is equivalent to the set of equations*

$$\left(\left(\frac{\alpha\pi}{R}\right)^2 - \gamma_{n,m}^2\right) \langle V', u_{m,n,j} \rangle_{\mathrm{L}^2(B)} = \langle G^u, u_{m,n,j} \rangle_{\mathrm{L}^2(B)} + \eta_m \delta_{n,0} \delta_{j,0}$$

for every $m \in \mathbb{N}$, $n \in \mathbb{N}_0$ and $j \in \{-n, \ldots, n\}$. Here the η_m are constants defined by the undetermined function η from Lemma 7.1.3 via

$$\eta_m := \sqrt{4\pi} \langle \eta, G_{m,0} \rangle_{\mathrm{L}^2_w(0,R)}.$$

The constant α is the same as in Theorem 5.1.3.

Proof. As stated above, we start from Equation (7.2), then V' satisfies

$$V' = \sum_{m=1}^{\infty} \sum_{n=0}^{\infty} \sum_{j=-n}^{n} \langle V', u_{m,n,j} \rangle_{\mathrm{L}^2(B)} u_{m,n,j}$$

in the sense of $\mathrm{L}^2(B)$. Since $\Delta V' = 4\pi G F'$ on B, where $F' \in \mathrm{C}^2(\overline{B}) \subseteq \mathrm{L}^2(B)$, we know that $\Delta V' \in \mathrm{L}^2(B)$. We can then develop this quantity into the basis $u_{m,n,j}$ as well and use Lemma 7.2.13 to get

$$\begin{aligned}\Delta V' &= \sum_{m=1}^{\infty}\sum_{n=0}^{\infty}\sum_{j=-n}^{n} \langle \Delta V', u_{m,n,j}\rangle_{\mathrm{L}^2(B)} u_{m,n,j} \\ &= -\sum_{m=1}^{\infty}\sum_{n=0}^{\infty}\sum_{j=-n}^{n} \gamma_{n,m}^2 \langle V', u_{m,n,j}\rangle_{\mathrm{L}^2(B)} u_{m,n,j},\end{aligned}$$

again in the sense of $\mathrm{L}^2(B)$. With this the left-hand side of Equation (7.2) becomes

$$\sum_{m=1}^{\infty}\sum_{n=0}^{\infty}\sum_{j=-n}^{n} \left(\left(\frac{\alpha\pi}{R}\right)^2 - \gamma_{n,m}^2\right) \langle V', u_{m,n,j}\rangle_{\mathrm{L}^2(B)} u_{m,n,j},$$

since $2\pi G/K = \alpha^2\pi^2/R^2$ (see the proof of Theorem 5.1.3). On the right-hand side, we know that $G^u \in \mathrm{C}(\overline{B}) \subseteq \mathrm{L}^2(B)$ as the antiderivative of an integrable function, see Remark 6.5. Then Equation (7.2) yields that η is also square-integrable over B. Using the expansion into the $u_{m,n,j}$, we have

$$\begin{aligned}G^u + \eta &= \sum_{m=1}^{\infty}\sum_{n=0}^{\infty}\sum_{j=-n}^{n} \langle G^u, u_{m,n,j}\rangle_{\mathrm{L}^2(B)} u_{m,n,j} \\ &\quad + \sum_{m=1}^{\infty}\sum_{n=0}^{\infty}\sum_{j=-n}^{n} \langle \eta, u_{m,n,j}\rangle_{\mathrm{L}^2(B)} u_{m,n,j}\end{aligned}$$

in $\mathrm{L}^2(B)$. Comparing coefficients produces the set of equations

$$\left(\left(\frac{\alpha\pi}{R}\right)^2 - \gamma_{n,m}^2\right) \langle V', u_{m,n,j}\rangle_{\mathrm{L}^2(B)} = \langle G^u, u_{m,n,j}\rangle_{\mathrm{L}^2(B)} + \langle \eta, u_{m,n,j}\rangle_{\mathrm{L}^2(B)}$$

for every valid index m, n and j. Now we just simplify the inner product of η with $u_{m,n,j}$. If $n \neq 0$, then

$$\begin{aligned}\langle \eta, u_{m,n,j}\rangle_{\mathrm{L}^2(B)} &= \int_0^R r^2 G_{m,n}(r)\eta(r) \int_{\mathbb{S}^2} Y_{n,j}(\xi)\,\mathrm{d}\omega(\xi)\,\mathrm{d}r \\ &= \sqrt{4\pi}\int_0^R r^2 G_{m,n}(r)\eta(r) \int_{\mathbb{S}^2} Y_{n,j}(\xi) Y_{0,0}(\xi)\,\mathrm{d}\omega(\xi)\,\mathrm{d}r = 0,\end{aligned}$$

because the integral over $\mathbb{S}^2$ vanishes. On the other hand if $n = j = 0$, we have

$$\begin{aligned}\langle \eta, u_{m,n,j}\rangle_{\mathrm{L}^2(B)} = \langle \eta, u_{m,0,0}\rangle_{\mathrm{L}^2(B)} &= \int_0^R r^2 G_{m,0}(r)\eta(r) \int_{\mathbb{S}^2} Y_{0,0}(\boldsymbol{\xi})\,\mathrm{d}\omega(\boldsymbol{\xi})\,\mathrm{d}r \\ &= \sqrt{4\pi} \int_0^R r^2 G_{m,0}(r)\eta(r)\,\mathrm{d}r = \sqrt{4\pi}\,\langle \eta, G_{m,0}\rangle_{\mathrm{L}^2_w(0,R)},\end{aligned}$$

proving this lemma. □

To proceed further and calculate V' we need to determine the coefficients $\langle V', u_{m,n,j}\rangle_{\mathrm{L}^2(B)}$ via Lemma 7.3.1. Obviously another assumption is necessary, namely

Assumption 7.3.2 In the following, we assume that the polytropic constant $K > 0$ (and thus $\alpha = \sqrt{2GR^2/(\pi K)}$) satisfies

$$\sqrt{\frac{2\pi G}{K}} = \alpha\frac{\pi}{R} \neq \gamma_{n,m}$$

for every $n \in \mathbb{N}_0$ and $m \in \mathbb{N}$.

Remark 7.3.3 Assumption 7.3.2 is especially important in the case $n = m = 1$. This is because $\gamma_{1,1}$ is the smallest positive zero of

$$x \mapsto \frac{\sin(xR)}{xR},$$

meaning that $\gamma_{1,1} = \pi/R$. For the simplest radially symmetric model we expect $\alpha \approx 1$ (see Remark 5.2.5), implying $\alpha\pi/R \approx \gamma_{1,1}$. Instead of Assumption 7.3.2, it would then also be reasonable to demand that the coefficient $\langle V', u_{1,1,j}\rangle_{\mathrm{L}^2(B)}$ vanishes, if α is so close to 1 that no other $\gamma_{n,m}$ can be equal to $\alpha\pi/R$.

If we just wanted to use V' or F' for calculating the exterior gravitational coefficients J_n (see Theorem 3.2.1), for example in an inverse problem, then the radial indeterminancy $\eta(r)$ would not matter (see Lemma 8.2.3). But if we want to uniquely determine V' and F', then we need an additional assumption, which we will specify now.

Theorem 7.3.4 *If the wind-induced potential V' satisfies Equation (7.2) in B and Assumption 7.3.2, then it is uniquely determined on this set under the additional assumption*

$$\int_{\mathbb{S}^2} V'(r\xi)\,\mathrm{d}\omega(\xi) = 0 \quad \textit{for every } r \in (0, R). \tag{7.8}$$

In this case V' is given as

$$V' = \sum_{m=1}^{\infty}\sum_{n=1}^{\infty}\sum_{j=-n}^{n} \frac{\langle G^u, u_{m,n,j}\rangle_{\mathrm{L}^2(B)}}{\left(\frac{\alpha\pi}{R}\right)^2 - \gamma_{n,m}^2} u_{m,n,j} \tag{7.9}$$

in the sense of $\mathrm{L}^2(B)$.

Proof. The functions $(G_{m,0})_{m\in\mathbb{N}}$ form an orthonormal basis of $\mathrm{L}^2_w(0, R)$. Thus Equation (7.8) is equivalent to

$$\int_0^R r^2 G_{m,0}(r) \int_{\mathbb{S}^2} V'(r\xi)\,\mathrm{d}\omega(\xi)\,\mathrm{d}r = 0 \quad \text{for every } m \in \mathbb{N}.$$

Since $Y_{0,0}$ is constant on $\mathbb{S}^2$ and $u_{m,0,0}(r\xi) = G_{m,0}(r)Y_{0,0}(\xi)$, we can rewrite this as

$$\langle V', u_{m,0,0}\rangle_{\mathrm{L}^2(B)} = 0 \quad \text{for every } m \in \mathbb{N}.$$

Now we can use Lemma 7.3.1 together with Assumption 7.3.2, which implies that

$$\langle V', u_{m,n,j}\rangle_{\mathrm{L}^2(B)} = \frac{\langle G^u, u_{m,n,j}\rangle_{\mathrm{L}^2(B)}}{\left(\frac{\alpha\pi}{R}\right)^2 - \gamma_{n,m}^2}$$

for every $m, n \in \mathbb{N}$ and $j \in \{-n, \ldots, n\}$. Combining this with

$$\langle V', u_{m,0,0}\rangle_{\mathrm{L}^2(B)} = 0$$

for every $m \in \mathbb{N}$, the function $V' \in \mathrm{L}^2(B)$ is then uniquely determined, since all of the coefficients $\langle V', u_{m,n,j}\rangle_{\mathrm{L}^2(B)}$ are known. □

There is one more relatively obvious simplification we can make. This is because V' satisfies Assumption 3.1.1, as all the physical quantities in this thesis do. This leads to the conclusion that in integrals of the form

$$\int_{\mathbb{S}^2} V'(r\xi)Y_{n,j}(\xi)\,\mathrm{d}\omega(\xi),$$

the integral over the φ-coordinate (in spherical coordinates) can be written as

$$\int_0^{2\pi} \sin(j\varphi)\,\mathrm{d}\varphi \quad \text{or} \quad \int_0^{2\pi} \cos(j\varphi)\,\mathrm{d}\varphi.$$

Both of these vanish if $j \neq 0$. Since the coefficients $\langle V', u_{m,n,j}\rangle_{\mathrm{L}^2(B)}$ involve integrals of this kind, they vanish if $j \neq 0$, so in the situation of Theorem 7.3.4 we have

$$V' = \sum_{m=1}^{\infty}\sum_{n=1}^{\infty}\sum_{j=-n}^{n} \frac{\langle G^u, u_{m,n,j}\rangle_{\mathrm{L}^2(B)}}{\left(\frac{\alpha\pi}{R}\right)^2 - \gamma_{n,m}^2} u_{m,n,j} = \sum_{m=1}^{\infty}\sum_{n=1}^{\infty} \frac{\langle G^u, u_{m,n,0}\rangle_{\mathrm{L}^2(B)}}{\left(\frac{\alpha\pi}{R}\right)^2 - \gamma_{n,m}^2} u_{m,n,0}.$$

Remark 7.3.5 Since $Y_{n,0}(\xi) = \sqrt{\frac{2n+1}{4\pi}} P_n(\xi_3)$ and $G_{m,n}(r) = c_{m,n} j_n(\gamma_{n,m} r)$, we have

$$\begin{aligned} u_{m,n,0}(x) = u_{m,n,0}(r\xi) &= \sqrt{\frac{2n+1}{4\pi}}\sqrt{\frac{2}{R^3 j_n^2(\gamma_{n,m}R)}} j_n(\gamma_{n,m}r)P_n(\xi_3) \\ &= \sqrt{\frac{n+1/2}{\pi R^3}}\frac{j_n(\gamma_{n,m}r)}{|j_n(\gamma_{n,m}R)|}P_n(\xi_3) \end{aligned}$$

for every $n, m \in \mathbb{N}$.

An Inverse Problem

8

In this chapter, we will see how to formulate the relationship between the wind field and the gravitational data as an inverse problem for some parameters of the wind field.

8.1 Structure of the Wind Field

We need to specify the structure of the zonal wind field u_φ a bit more, to be able to determine some relevant physical parameters of it. One way to do this, which is common in the literature [10, 11] but presents some problems from a mathematical point of view (see Remark 8.1.4), will be defined now. These models are based on projecting the wind field on the surface of the planet, which is known from direct observation (for Saturn and Jupiter see [4] and [17] respectively), into the planet along some axis of symmetry. To illustrate the surface wind, some pictures and graphs are helpful, see Fig. 8.1. We will refer to the surface wind field as a function of $t = \cos(\theta)$ (in spherical coordinates) as $u^{\mathrm{surf}} : [-1, 1] \to \mathbb{R}, \quad t \mapsto u^{\mathrm{surf}}(t)$.

Definition 8.1.1 We define the **cylindrically projected wind** as the function

$$u_{\mathrm{c}}^{\mathrm{proj}} : B_R(0) \setminus \{0\} \to \mathbb{R}, \quad r\xi \mapsto u^{\mathrm{surf}}\left(\mathrm{sign}(\xi_3)\sqrt{1 - \frac{r^2}{R^2}(1 - \xi_3^2)}\right).$$

T. Peter, *Reconstructing Wind Fields from Gravitational Data on Gas Giants*, BestMasters, https://doi.org/10.1007/978-3-658-51277-4_8

Jupiter as seen by Cassini

Image credit: NASA/JPL/University of Arizona

Saturn as seen by Cassini

Image credit: NASA/JPL/Space Science Institute

Both pictures above are from `https://photojournal.jpl.nasa.gov/`

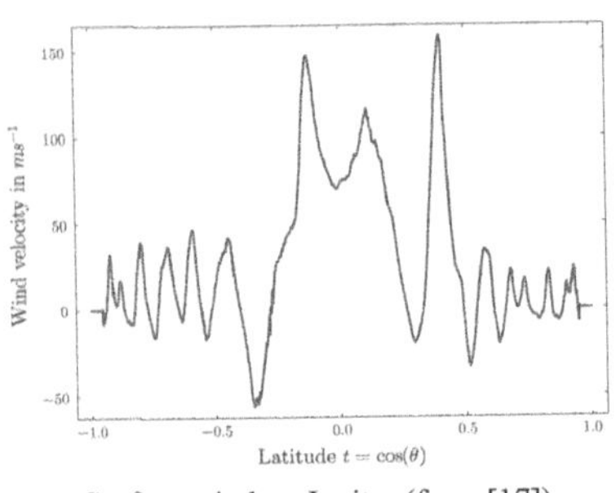

Surface wind on Jupiter (from [17])

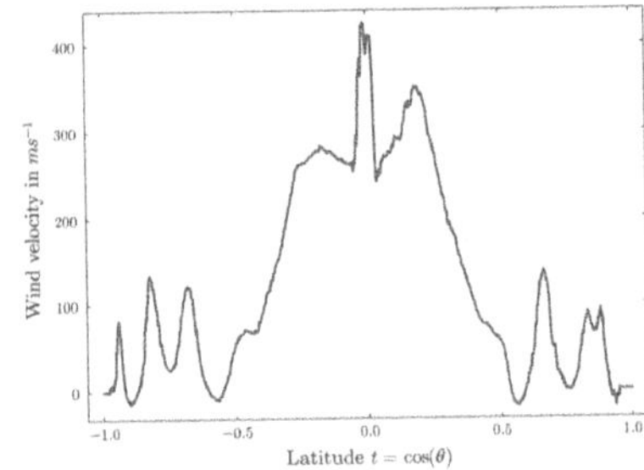

Surface wind on Saturn (from [4])

Fig. 8.1 The surface wind on Jupiter and Saturn visible from space and as functions of the latitude $t = \cos(\theta)$

In spherical coordinates this would be

$$u_{\mathrm{c}}^{\mathrm{proj}}(r\boldsymbol{\xi}(\varphi, t)) = u^{\mathrm{surf}}\left(\operatorname{sign}(t)\sqrt{1 - \frac{r^2}{R^2}(1 - t^2)}\right).$$

Definition 8.1.2 We define the spherically or **radially projected wind** as the function

$$u_{\mathrm{s}}^{\mathrm{proj}} : B_R(0) \setminus \{0\} \to \mathbb{R}, \quad r\boldsymbol{\xi} \mapsto u^{\mathrm{surf}}(\xi_3)$$

or in spherical coordinates

$$u_{\mathrm{s}}^{\mathrm{proj}}(r\boldsymbol{\xi}(\varphi, t)) = u^{\mathrm{surf}}(t).$$

Remark 8.1.3 The definition of u_c^{proj} is justified by the geometric consideration in Fig. 8.2. We imagine the planetary surface as a sphere of radius R around 0 and use spherical coordinates (r, φ, t). We know the wind on the surface, which is given by $u^{surf}(t)$. Since the quantities involved are independent of φ, the problem becomes two-dimensional, so if we want to determine the value of $u_c^{proj}(r\boldsymbol{\xi}(\varphi, t))$ at some interior point $(r, \cos(\theta))$ we continue along a line parallel to the $\boldsymbol{\varepsilon}^3$-axis until we reach the closest point on the sphere. This point will have coordinates $(R, \cos(\theta'))$. Then we demand

$$u_c^{proj}(r\boldsymbol{\xi}(\varphi, t)) = u^{surf}(\cos(\theta')).$$

We need a relationship between $(r, \cos(\theta))$ and $\cos(\theta')$, so we define the distance between the $\boldsymbol{\varepsilon}^3$–axis (also the rotation axis) and the line through the points defined by $(r, \cos(\theta))$ and $(R, \cos(\theta'))$ as $d \geq 0$. Then by the Pythagorean theorem

$$R^2 = d^2 + R^2\cos^2(\theta') \quad \text{and} \quad r^2 = d^2 + r^2\cos^2(\theta),$$

which implies $\cos^2(\theta') = 1 - (r/R)^2(1 - \cos^2(\theta))$. Depending on the sign of $\cos(\theta)$, the closest point on the sphere is either above or below the xy-plane, so if we substitute $t = \cos(\theta)$ and $t' = \cos(\theta')$, then

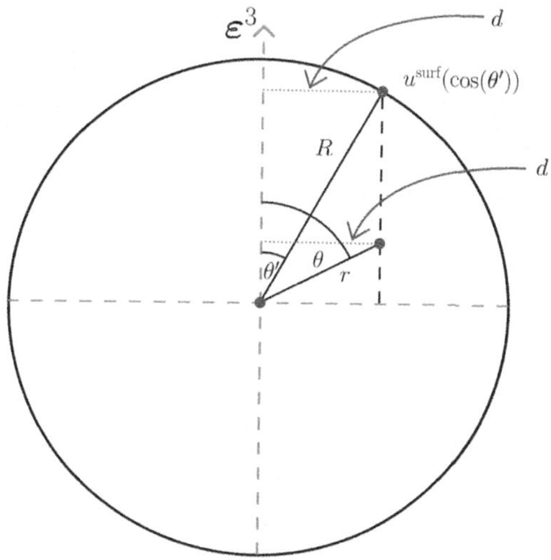

Fig. 8.2 Calculating u_c^{proj} from u^{surf}

$$t' = \operatorname{sign}(t)\sqrt{1 - \frac{r^2}{R^2}(1 - t^2)}.$$

This justifies the definition of u_c^{proj}.

Remark 8.1.4 The definition of the cylindrically projected wind presents a mathematical problem, namely a discontinuity at the equatorial plane. This arises because the surface wind field is not symmetrical with respect to the equator. Thus projecting the wind field inward parallel to the axis of rotation leads to different values of u_c^{proj} when approaching the equatorial plane from the North and the South. This discontinuity is illustrated in Figs. 8.3a and 8.3b in the plane $x_2 = 0$. This fact contradicts our assumption that $\boldsymbol{u} \in C^2(\overline{B}, \mathbb{R}^3)$, which we made at the beginning of Chap. 4. Since the formulas we derived for calculating F' (or equivalently V') all include a term which includes a spatial derivative of $\boldsymbol{u}$ or $\boldsymbol{\varepsilon}^{\varphi} \cdot \boldsymbol{u} = u_{\varphi}$, see Equations (6.1) and (7.1), this is especially problematic. These facts have also been discussed in the literature [12] and the discontinuity will contribute a significant amount to the calculation of the coefficients J_n even when some smoothing is applied along the discontinuity (also [12]). Therefore this approach is less coherent from a mathematical point of view.

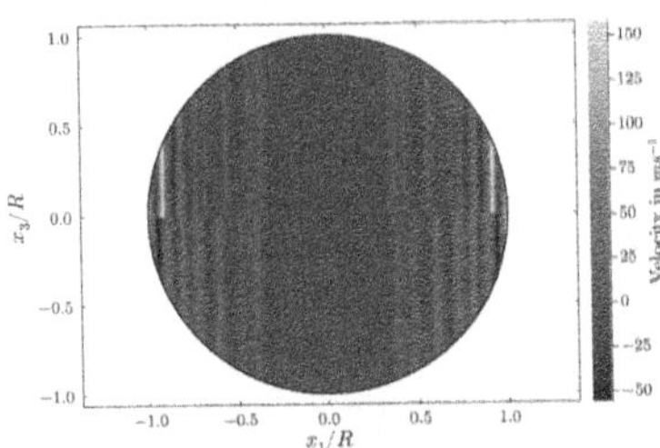

(a) Cylindrically projected wind for the planet Jupiter

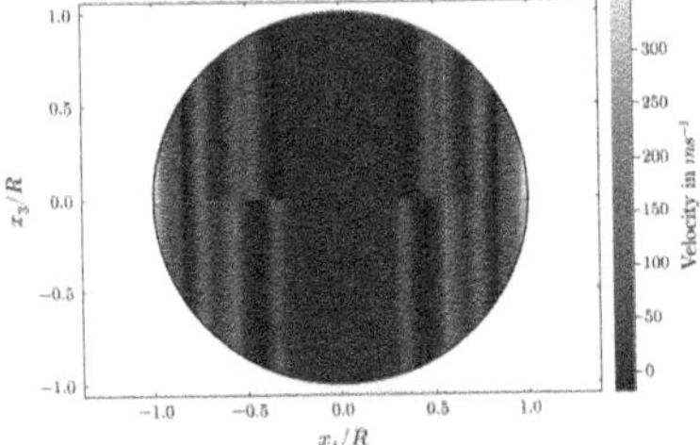

(b) Cylindrically projected wind for the planet Saturn

Fig. 8.3 Both plots show the plane $x_2 = 0$ of the planet. The axes are scaled with $1/R$. The data is taken from [17] for Jupiter and [4] for Saturn

The projected wind fields are then usually multiplied by a function Q_p, which characterizes the decrease of the wind field in the interior of the planet (see [10, 11]). These functions are dependent on parameters $p \in D \subseteq \mathbb{R}^d$ and can have the following forms.

Example 8.1.5 Let $D := (0, 1]$ and for every $p \in D$

$$Q_p : B_R(0) \to [0, 1], \quad \boldsymbol{x} \mapsto \exp\left(\frac{|\boldsymbol{x}|/R - 1}{p}\right).$$

Example 8.1.6 Let $D := (0, 1] \times [0, 1] \times \mathbb{R}^+$ and for every $p \in D$

$$Q_p : B_R(0) \to [0, 1], \quad \boldsymbol{x} \mapsto p_2 \exp\left(\frac{|\boldsymbol{x}|/R - 1}{p_1}\right) + (1 - p_2) \frac{\tanh\left(\frac{|\boldsymbol{x}|/R + p_1 - 1}{p_3}\right) + 1}{\tanh\left(\frac{p_1}{p_3}\right) + 1}.$$

These functions all satisfy $Q_p(\boldsymbol{x}) = 1$ for all $p \in D$ and $\boldsymbol{x} \in \partial B_R(0)$.
To summarize, the azimuthal component of the wind field u_φ can be decomposed as

$$u_\varphi(\boldsymbol{x}) = Q_p(\boldsymbol{x}) u^{\mathrm{proj}}(\boldsymbol{x})$$

where u^{proj} is one of the projected wind fields from above and Q_p is a function dependent on parameters $p \in D$, satisfying

$$\begin{aligned} Q_p(\boldsymbol{x}) &= 1 \quad \text{for every } \boldsymbol{x} \in \partial B_R(0) \text{ and} \\ 0 \leq Q_p(\boldsymbol{x}) &\leq 1 \quad \text{for every } \boldsymbol{x} \in B_R(0). \end{aligned}$$

8.2 An Inverse Problem Involving the Wind-Induced Gravitational Coefficients

We will first recapitulate the definition of the gravitational coefficients J_n, which first occurred in Theorem 3.2.1.

Definition 8.2.1 For a mass density $F \in C^2(\overline{B_R(0)})$ satisfying Assumption 3.1.1 we define the (tesseral) **gravitational coefficients** caused by F as

$$J_n := -\frac{1}{MR^n} \int_{B_R(0)} |x|^n P_n\left(\frac{x_3}{|x|}\right) F(x)\,dx$$

for every $n \in \mathbb{N}$ with $n \geq 2$.

Since these coefficients first occurred, we have split up the density F into two terms F_0 and F', corresponding to a perturbation approach. We can thus split up the J_n in the same way.

Definition 8.2.2 For the mass density $F_0 \in C^2(\overline{B_R(0)})$ from Definition 4.2.1 we define the **static gravitational coefficients** as

$$J_n^0 := -\frac{1}{MR^n} \int_{B_R(0)} |x|^n P_n\left(\frac{x_3}{|x|}\right) F_0(x)\,dx.$$

For the dynamic or wind-induced density $F' \in C^2(\overline{B_R(0)})$ from Definition 4.2.2 we define the **dynamic gravitational coefficients** as

$$\delta J_n := -\frac{1}{MR^n} \int_{B_R(0)} |x|^n P_n\left(\frac{x_3}{|x|}\right) F'(x)\,dx.$$

For both of these terms $n \in \mathbb{N}$ with $n \geq 2$.

One important property of these coefficients is their invariance under the addition of purely radial functions.

Lemma 8.2.3 *If the density $F \in C^2(\overline{B_R(0)})$ is a radial function, meaning that $F(x) = f(|x|)$ for all $x \in B_R(0)$ and some $f \in C^2(\mathbb{R}_0^+)$, then*

$$J_n = 0 \quad \textit{for all } n \in \mathbb{N} \textit{ and } n \geq 2.$$

Proof. For any $n \in \mathbb{N}$ with $n \geq 2$ we use the decomposition $x = r\xi$ introduced in the first chapter and get

$$J_n = -\frac{1}{MR^n}\int_0^R \int_{\mathbb{S}^2} r^{n+2} P_n(\xi_3) f(r)\,\mathrm{d}\omega(\boldsymbol{\xi})\,\mathrm{d}r$$
$$= -\frac{1}{MR^n}\int_0^R r^{n+2} f(r)\,\mathrm{d}r \int_{\mathbb{S}^2} P_n(\xi_3)\,\mathrm{d}\omega(\boldsymbol{\xi}).$$

We use spherical coordinates to further simplify the second integral, so

$$\int_{\mathbb{S}^2} P_n(\xi_3) = \int_0^{2\pi}\int_{-1}^{1} P_n(t)\,\mathrm{d}t\,\mathrm{d}\varphi$$
$$= 2\pi \int_{-1}^{1} P_n(t)\,\mathrm{d}t = 4\pi\,\delta_{n0},$$

since the Legendre polynomials are orthogonal and $P_0(t) = 1$. Finally, since we assumed that $n \geq 2$, $\delta_{n,0} = 0$ and thus J_n must vanish as well. □

As stated in previous chapters, the gravitational coefficients J_n of the gas giants Jupiter and Saturn have been measured by the Juno and the Cassini missions respectively and are known up to J_{10}, see [8, 9, 11] (more recently these coefficients have been determined up to J_{40}, see [10]).

Definition 8.2.4 For a natural number $N \geq 2$ we define $J := (J_n)_{n=2,\dots,N} \in \mathbb{R}^{N-1}$ as the **measured gravitational coefficients** (corresponding to the first columns in Tables 8.1 and 8.2).

Table 8.1 The gravitational coefficients of Jupiter. The first column represents the values measured by Juno. The second column contains averages from an ensemble of interior models, see [11]

Entries $\times 10^{-6}$	Measured value	Interior model value
J_2	14696.572	14696.572
J_3	−0.042	0.0
J_4	−586.609	−586.609
J_5	−0.069	0.0
J_6	34.198	34.188
J_7	0.124	0.0
J_8	−2.426	−2.461
J_9	−0.106	0.0
J_{10}	0.172	0.202

Table 8.2 The gravitational coefficients of Saturn. The first column represents the values measured by Cassini. The second column contains averages from an ensemble of interior models, see [11]

Entries $\times 10^{-6}$	Measured value	Interior model value
J_2	16290.573	16300.0
J_3	0.059	0.0
J_4	−935.314	−925.76
J_5	−0.224	0.0
J_6	86.34	82.326
J_7	0.108	0.0
J_8	−14.624	−9.226
J_9	0.369	0.0
J_{10}	4.672	1.188

The vector $J^0 := (J_n^0)_{n=2,\dots,N} \in \mathbb{R}^{N-1}$ contains the **gravitational coefficients of the interior models** (corresponding to the second columns in Tables 8.1 and 8.2).

In the previous chapters we have seen how to calculate the wind-induced density F' from the zonal wind field u_φ, with an indeterminancy consisting of an additive function η which only depends on the radial coordinate r. By Lemma 8.2.3, this means we can determine the coefficients δJ_n uniquely if we know the wind field. As we have discussed, the wind field is dependent on a vector of parameters $p \in \mathbb{R}^d$ and the coefficients δJ_n depend on the wind field. We will write $\delta J_n = \delta J_n(p)$ to emphasize this dependency.

Definition 8.2.5 Our **inverse problem** is this. Given the vectors J, $J^0 \in \mathbb{R}^{N-1}$ for some natural number $N \geq 2$, we define the vector field

$$\delta J : D \to \mathbb{R}^{N-1}, \quad p \mapsto (\delta J_n(p))_{n=2,\dots,N},$$

where $D \subseteq \mathbb{R}^d$ is the set of all valid parameters p. Then we want to find the minimum of the functional

$$\mathcal{L} : D \to \mathbb{R}_0^+, \quad p \mapsto (J - (J^0 + \delta J(p)))^T W (J - (J^0 + \delta J(p))),$$

where $W \in \mathbb{R}^{(N-1)\times(N-1)}$ is a fixed symmetric and positive definite matrix. Often $W = I$ is chosen, such that $\mathcal{L}(p) = \|J - (J^0 + \delta J(p))\|_2^2$.

9 Conclusion

This thesis provides a mathematically rigorous framework for the development of solutions of two models (those examined in Chaps. 6 and 7), which describe the relationship between the gravitational potential and the wind field in the atmosphere of a gas giant, like Jupiter or Saturn. It also provides some simple criteria which, when applied, provide unique wind-induced densities or potentials in the interior of the planet.

That being said, this thesis is far from a complete overview of the subject, so we will now discuss some shortcomings of this work. One problem is the choice of static model. We assumed that our model was spherically symmetric, meaning we could describe our planet as a ball of radius $R > 0$. This is also done for Earth, for example in the well known preliminary reference Earth model PREM from [3], but the gas giants Jupiter and Saturn deviate from a spherical shape far more than Earth. In fact, while Earth's ellipticity ($1 - b/a$, where a is the equatorial and b is the polar radius) is 0.00335, for Jupiter and Saturn these values are at 0.06487 and 0.09796 [20, 21]. Since these numbers are an order of magnitude greater than Earth's, we need to be careful about applying a spherically symmetric model.

The other choice we made was to use a polytrope of index unity to describe the relation between pressure and density. This choice allowed us to express the static density as a smooth function, which is advantageous from a mathematical point of view. But there are other approaches, such as the CMS model (described in [7]), which uses a series of concentric Maclaurin spheroids of constant density. The polytropic model can also be used to generate a non-spherical planetary boundary, this is described in [22]. These two methods have been compared in [23] as well.

T. Peter, *Reconstructing Wind Fields from Gravitational Data on Gas Giants*, BestMasters, https://doi.org/10.1007/978-3-658-51277-4_9

It is worth noting that the uniqueness condition we derived in Theorem 7.3.4 has no physical justification so far. Another problem we have discussed previously is the nature of the function $u_{\mathrm{c}}^{\mathrm{proj}}$, defined in Definition 8.1.1. This function has a jump discontinuity at the equatorial plane, which is inconsistent with the previous modelling, which assumed that the wind field is twice continuously differentiable. This problem is further discussed in [12] and [2].

Nonetheless this thesis provides a structured and detailed introduction to the subject of inverse gravimetry on gas giants for the purpose of determining atmospheric structure.

References

1. Abramowitz, M., Stegun, I.A.: Handbook of Mathematical Functions: With Formulas, Graphs and Mathematical Tables, 10th ed. Washington, D.C.: U.S. Government Printing Office (1972)
2. Dietrich, W., Wulff, P., Wicht, J., Christensen, U.: Linking zonal winds and gravity—ii. explaining the equatorially antisymmetric gravity moments of jupiter. Monthly Notices of the Royal Astronomical Society **505**, 3177–3191 (2021)
3. Dziewonski, A., Anderson, D.: Preliminary reference earth model. Physics of the Earth and Planetary Interiors **25**, 297–356 (1981)
4. García-Melendo, E., Pérez-Hoyos, S., Sánchez-Lavega, A., Hueso, R.: Saturn's zonal wind profile in 2004–2009 from cassini iss images and its long-term variability. Icarus **215**, 62–74 (2011)
5. Heuser, H.: Lehrbuch der Analysis: Teil 1, 14 edn. B.G. Teubner Verlag (2001)
6. Hubbard, W.: Gravitational signature of jupiter's deep zonal flows. Icarus **137**, 357–359 (1998)
7. Hubbard, W.: Concentric maclaurin spheroid models of rotating liquid planets. The Astrophysical Journal **768** (2013). Article number 43
8. Iess, L., Folkner, W., Durante, D., Parisi, M., Kaspi, Y., Galanti, E., Guillot, T., Hubbard, W., Stevenson, D., Anderson, J., Buccino, D., Casajus, L., Milani, A., Park, R., Racioppa, P., Serra, D., Tortora, P., Zannoni, M., Cao, H., Helled, R., Lunine, J., Miguel, Y., Militzer, B., Wahl, S., Connerney, J., Levin, S., Bolton, S.: Measurement of jupiter's asymmetric gravity field. Nature **555**, 220–222 (2018)
9. Iess, L., Militzer, B., Kaspi, Y., Nicholson, P., Durante, D., Racioppa, P., Anabtawi, A., Galanti, E., Hubbard, W., Mariani, M., Tortora, P., Wahl, S., Zannoni, M.: Measurement and implications of saturn's gravity field and ring mass. Science **364** (2019). Page identifier eaat2965
10. Kaspi, Y., Galanti, E., Park, R., Duer, K., Gavriel, N., Durante, D., Iess, L., Parisi, M., Buccino, D., Guillot, T., Stevenson, D., Bolton, S.: Observational evidence for cylindrically oriented zonal flows on jupiter. Nat Astron **7**, 1463–1472 (2023)
11. Kaspi, Y., Galanti, E., Showman, A., Stevenson, D., Guillot, T., Iess, L., Bolton, S.: Comparison of the deep atmospheric dynamics of jupiter and saturn in light of the juno and cassini gravity measurements. Space Sci Rev **216** (2020). Article number 84

T. Peter, *Reconstructing Wind Fields from Gravitational Data on Gas Giants*, BestMasters, https://doi.org/10.1007/978-3-658-51277-4

12. Kong, D., Zhang, K., Schubert, G.: Odd gravitational harmonics of jupiter: Effects of spherical versus nonspherical geometry and mathematical smoothing of the equatorially antisymmetric zonal winds across the equatorial plane. Icarus **277**, 416–423 (2016)
13. Michel, V.: Lectures on Constructive Approximation: Fourier, Spline, and Wavelet Methods on the Real Line, the Sphere, and the Ball, 1 edn. Birkhäuser (2013)
14. Michel, V.: Geomathematics: Modelling and Solving Mathematical Problems in Geodesy and Geophysics, 1 edn. Cambridge University Press (2022)
15. Michel, V., Fokas, A.: A unified approach to various techniques for the non-uniqueness of the inverse gravimetric problem and wavelet-based methods. Inverse Problems **24** (2008). Article number 045019
16. Müller, C.: Foundations of the Mathematical Theory of Electromagnetic Waves, 1 edn. Springer Berlin (1969)
17. Tollefson, J., Wong, M., de Pater, I., Simon, A., Orton, G., Rogers, J., Atreya, S., Cosentino, R., Januszewski, W., Morales-Juberías, R., Marcus, P.: Changes in jupiter's zonal wind profile preceding and during the juno mission. Icarus **296**, 163–178 (2017)
18. Watson, G.: A Treatise on the Theory of Bessel Functions, 2 edn. Cambridge University Press (1980)
19. Wicht, J., Dietrich, W., Wulff, P., Christensen, U.: Linking zonal winds and gravity: The relative importance of dynamic self gravity. Monthly Notices of the Royal Astronomical Society **492**, 3364–3374 (2020)
20. Williams, D.: Jupiter fact sheet. https://nssdc.gsfc.nasa.gov/planetary/factsheet/jupiterfact.html (2024). Accessed: June 06, 2024 at 11:46am
21. Williams, D.: Saturn fact sheet. https://nssdc.gsfc.nasa.gov/planetary/factsheet/saturnfact.html (2024). Accessed: June 06, 2024 at 11:47am
22. Wisdom, J.: Non-perturbative hydrostatic equilibrium. http://web.mit.edu/wisdom/www/interior.pdf (1996). Accessed: May 22, 2024 at 8:57am
23. Wisdom, J., Hubbard, W.: Differential rotation in jupiter: A comparison of methods. Icarus **267**, 315–322 (2016)

The manufacturer's authorised representative in the EU is Springer Nature Customer Service Centre GmbH, Europaplatz 3, 69115 Heidelberg, Germany. If you have any concerns regarding our products, please contact ProductSafety@springernature.com

Printed and bound by CPI Group (UK) Ltd, Croydon, CR0 4YY

07/07/2026

02160931-0001